AF510403

Dynamic Fair Gain-Sharing Cooperation

Theory, Solution Techniques and
A Reference Solution for United Nations
Cooperation Initiatives

Dynamic Fair Gain-Sharing Cooperation

Theory, Solution Techniques and
A Reference Solution for United Nations
Cooperation Initiatives

David W. K. Yeung

Hong Kong Shue Yan University, Hong Kong
St. Petersburg State University, Russia

Leon A Petrosyan

St. Petersburg State University, Russia

World Scientific

EW JERSEY • LONDON • SINGAPORE • BEIJING • SHANGHAI • HONG KONG • TAIPEI • CHENNAI • TOKYO

Published by

World Scientific Publishing Europe Ltd.
57 Shelton Street, Covent Garden, London WC2H 9HE
Head office: 5 Toh Tuck Link, Singapore 596224
USA office: 27 Warren Street, Suite 401-402, Hackensack, NJ 07601

Library of Congress Control Number: 2026022888

British Library Cataloguing-in-Publication Data
A catalogue record for this book is available from the British Library.

DYNAMIC FAIR GAIN-SHARING COOPERATION
Theory, Solution Techniques and A Reference Solution for United Nations Cooperation Initiatives

Copyright © 2026 by World Scientific Publishing Europe Ltd.

All rights reserved. This book, or parts thereof, may not be reproduced in any form or by any means, electronic or mechanical, including photocopying, recording or any information storage and retrieval system now known or to be invented, without written permission from the publisher.

For photocopying of material in this volume, please pay a copying fee through the Copyright Clearance Center, Inc., 222 Rosewood Drive, Danvers, MA 01923, USA. In this case permission to photocopy is not required from the publisher.

ISBN 978-1-80061-954-8 (hardcover)
ISBN 978-1-80061-955-5 (ebook for institutions)
ISBN 978-1-80061-956-2 (ebook for individuals)

For any available supplementary material, please visit
https://www.worldscientific.com/worldscibooks/10.1142/Q0574#t=suppl

Desk Editors: Murali Appadurai/Pui Yee Lum

Typeset by Stallion Press
Email: enquiries@stallionpress.com

Preface

It is widely recognized that non-cooperative behaviors among participants in an interactive encounter generally result in outcomes that are not Pareto-optimal. Cooperation offers the best promise to achieve a group-optimal solution. However, despite the advantages of cooperation, sustainable collaborations encounter significant challenges related to individual rationality, group optimality, time consistency, and a gain-sharing mechanism that is deemed acceptable by all participants. An appropriate gain-sharing principle is essential for securing the agreement of all participants to collaborate and guarantee the sustainability of the scheme. Numerous instances of collaborative scheme failures have been attributed to dissatisfaction with the gain-sharing arrangements. Gain-sharing arrangements are a vital component in establishing sustainable cooperation. In the absence of a proper gain-sharing mechanism, certain participants may receive significantly less benefit compared to their contributions to other participants under cooperation.

The asymmetry of participants regarding size, functionality, capability, resources, facilities, and growth potential poses significant difficulties in designing gain-sharing arrangements that are acceptable and perceived as fair by all involved parties. A fair gain-sharing principle for cooperative gains is essential for securing the consensus of all players to collaborate and ensuring the sustainability of the scheme. To achieve fair gain-sharing, the distribution of cooperative gains has to reflect the contributions made by the players. In this book, we present a new technique to measure the contributions of the participants to the group-optimal payoff and derive a dynamic fair gain-sharing solution that reflects the contributions of the participants. The solution entails individual rationality, group optimality, time consistency, and the fair gain-sharing principle. This novel analysis provides a new element in dynamic cooperative solutions that is rooted in economic rationality and perceptions of fairness. It is expected to offer an effective means to resolve real-life challenges in collaboration and to stand as a reference solution to the United Nations' cooperation initiatives that strongly emphasize fairness.

Our gratitude to Dr. Yingxuan Zhang for her excellent technical assistance in compiling this book. We are grateful to Nina and Ovanes (LP) and Stella and Patricia (DY) for their love and patience during this book project, which at times may have diverted our focus from them.

David W. K. Yeung
Leon A. Petrosyan

About the Authors

David W. K. Yeung is Director of SRS Consortium for Advanced Study in Cooperative Dynamic Games at Shue Yan University and Co-director of the Centre of Game Theory at Saint Petersburg State University. Yeung obtained his PhD in mathematical economics from York University. He studied in the Doctor of Sciences (Doktor Nauk — Continental European higher doctorate above PhD) program in applied mathematics at Saint Petersburg State University and was awarded the University's highest degree of Dr.h.c. for his outstanding contributions in differential games. He has held faculty positions at Queen's University, the University of Hong Kong, and the National University of Singapore.

Yeung's main areas of research include game theory, optimization, operations research, and stochastic processes, and he has published extensively in these areas. He serves as Managing Editor of *International Game Theory Review*, Guest Editor of *Annals of Operations Research*, *Journal of the Operations Research Society of China*, and *Computational and Applied Mathematics*, Associate Editor of *Dynamic Games & Applications*, and Associate Editor of *Operations Research Letters*. He and Leon Petrosyan have co-authored the world's first books on *Cooperative Stochastic Differential Games (Springer)*, *Subgame Consistent Economic Optimization (Birkhauser)*, *Subgame Consistent Cooperation (Springer)*, and *Durable-strategies Dynamic Games (Springer)*.

Leon A. Petrosyan is Professor and Head of the Department of Mathematical Game Theory and Statistical Decisions and Dean of the Faculty of Applied Mathematics-Control Processes at St. Petersburg State University. Petrosyan obtained his Ph.D. at Vilnius State University, Master's and Dr. of Sci. at St. Petersburg State University. In 2008, he was elected as the President of the International Society of Dynamic Games (ISDG) and a foreign member of the Academy of Sciences of the Republic of Armenia. He was also awarded the Issacs Award, the highest prize in dynamic games, in 2014.

Petrosyan's main areas of research include theory of games. control theory, operations research, cooperative games, differential games, and mathematical ecology, and he has published extensively in these areas. He serves as the Editor of *International Game Theory Review*, *Game Theory and Applications*, and Chief Editor of *Mathematical Game Theory and Applications* and *Viestnik Petrburgskogo Universiteta* (*Applied Mathematics, Control, Informatics*). He and David Yeung have co-authored the world's first books on *Cooperative Stochastic Differential Games (Springer)*, *Subgame Consistent Economic Optimization (Birkhauser)*, *Subgame Consistent Cooperation (Springer)*, and *Durable-strategies Dynamic Games (Springer)*.

Contents

Preface v

About the Authors vii

1. Introduction **1**

2. Dynamic Fair Gain-Sharing Cooperation: Theory and Solution Technique **9**

1. The Notion of Fair Gain-Sharing 10
   1.1. Existing Gain-Sharing Schemes in Dynamic Games 10
   1.2. Fair Gain-Sharing 12
   1.3. Need for Fair Gain-Sharing: Examples 13
2. Basic Framework for Analysis 14
   2.1. Game Formulation 14
   2.2. Non-cooperative Equilibrium 15
   2.3. Group Optimal Outcome 16
3. Fair Gain-Sharing Cooperation 17
   3.1. Marginal Contributions of Players 18
   3.2. Fair Gain-Sharing Weight 20
4. A Dynamic Fair-Sharing Cooperative Solution 20
   4.1. A Dynamically Stable Fair Gain-Sharing Value 21
   4.2. Time-Consistent Payoff Distribution 23
   4.3. Time-Consistent versus Fully Time-Consistent 25

5. An Illustrative Example in Public Capital Development . . . 26
5.1. Non-cooperative Equilibrium . . . 27
5.2. Group Optimal Outcome . . . 28
5.3. Fair Gain-Sharing Weight . . . 31
5.4. A Dynamically Stable Fair Gain-Sharing Solution . . . 34
6. Concluding Remarks . . . 34
Appendices . . . 36
Appendix A. Proof of Proposition 5.1 . . . 36
Appendix B. Proof of Proposition 5.2 . . . 38
Appendix C. Proof of Proposition 5.3 . . . 40
References . . . 44

3. Application of Fair Gain-Sharing in Global Environmental Management 47

1. Gain-Sharing Problems in Global Environmental Management . . . 48
1.1. Asymmetries of the World Nations . . . 49
1.2. Conference of the Parties: A Review . . . 51
2. Theoretical Underpinnings of a Reference Solution . . . 59
3. Game Formulation and Equilibria of a Reference Solution . . . 61
3.1. Nations' Strategies, State Variables, and Evolution Dynamics . . . 61
3.2. Objectives and Payoffs . . . 62
3.3. Non-cooperative Equilibria . . . 62
3.4. Group Optimal Solution . . . 63
4. A Dynamic Fair-Sharing Cooperative Solution . . . 64
4.1. Marginal Contributions of Nations . . . 65
4.2. Fair Gain-Sharing Solution . . . 66
5. Explication in COP Resolution Plan . . . 67
6. A Numerical Example of International Climate Control . . . 69
6.1. Non-cooperative Equilibrium . . . 70
6.2. Group Optimal Outcome . . . 71
6.3. Fair Gain-Sharing Weight . . . 72
6.4. A Dynamically Stable Fair Gain-Sharing Value . . . 76

7. Concluding Remarks 76
Appendices 77
Appendix A. Proof of Proposition 6.1 77
Appendix B. Proof of Proposition 6.2 80
References 84

4. Dynamic Joint Ventures and Fair Gain-Sharing 85

1. Joint Ventures and Globalization 86
2. General Game Framework of Joint Venture 89
2.1. Basic Game Formulation and Benchmark Non-cooperative Equilibria 89
2.2. Venture Payoff Structure and Maximized Payoff 90
3. Fair Gain-sharing in Joint Venture 92
3.1. Marginal Contributions of Agents 93
3.2. Fair Gain-sharing Weight 95
4. A Dynamic Fair-sharing Cooperative Solution 95
4.1. A Time-consistent Fair Gain-sharing Value . . . 96
4.2. Time-consistent Payoff Distribution 97
5. An Illustrative Example 98
6. Concluding Remarks 105
Appendices 106
Appendix A. Proof of Proposition 5.1 106
Appendix B. Proof of Proposition 5.2 108
Appendix C. Proof of Proposition 5.3 110
References 113

5. Fair Gain-Sharing Under Random Exit 117

1. Random-horizon Dynamic Games 118
1.1. Game Formulation 119
1.2. Dynamic Programming for Random-horizon Optimization 119
2. Non-cooperative Equilibria and Group Optimality under Random Horizon 123
2.1. Non-cooperative Outcome 123
2.2. Group Optimality under Random Horizon 124
3. Fair Gain-sharing Cooperation under Random Exit 126
3.1. Marginal Contributions of Players 126

3.2. Fair Gain-sharing Solution 128
4. Subgame-consistent Solutions and Payment Mechanism 129
5. An Illustration in Fair Sharing under Random Exit 132
6. Concluding Remarks 139
Appendices 140
Appendix A. Proof of Proposition 5.1 140
Appendix B. Proof of Proposition 5.2 142
References 144

6. Dynamic Fair Gain-Sharing in Games on a Network 147

1. Networks 148
2. A Dynamic Network Game 149
2.1. Efficient Collaboration and Pareto Optimum 150
2.2. Marginal Contribution of Player in a Network 151
3. Fair Gain-sharing Solutions and Payoff Distribution 153
3.1. A Dynamic Fair Gain-sharing Solution 154
3.2. Time-consistent Payoff Distribution Procedure 155
4. An Illustrative Example in Interconnected Electricity Network 156
4.1. Joint Payoff Maximization under Collaboration 157
4.2. Maximized Joint Payoff without Company i . . . 158
4.3. Dynamically Stable Fair Gain-sharing Solution and PDP 161
5. Concluding Remarks 162
Appendices 162
Appendix A. Proof of Proposition 4.1 162
Appendix B. Proof of Propositions 4.2 and 4.3 166
References 172

7. Fair Gain-Sharing Under Dynamic Uncertainty 175

1. Stochastic Dynamic Games 176
1.1. Examples of Stochastic Dynamics 176

1.2. Game Formulation 177
1.3. Non-cooperative Equilibria 178
1.4. Group Optimality 179
2. Fair Gain-Sharing Cooperation 180
2.1. Marginal Contributions of Players 180
2.2. Fair Gain-Sharing Weight and Cooperative Solution 182
2.3. Subgame Consistent Solutions and Payment Mechanism 183
3. An Illustration in Cooperative Resource Extraction . . 185
4. A Note on Uncertain Future Payoff Structures 190
5. Concluding Remarks 192
Appendices 193
Appendix A. Proof of Proposition 3.1 193
Appendix B. Proof of Proposition 3.2 195
References 198

Index 201

Chapter 1

Introduction

It is widely recognized that non-cooperative behaviors among participants generally result in outcomes that are not Pareto-optimal. Even more concerning, highly unfavorable outcomes (such as the prisoner's dilemma) and potentially catastrophic results (like the tragedy of the commons) may arise when the parties involved prioritize their individual self-interests in a non-cooperative context. In a constantly changing world, non-cooperative behaviors driven by short-sighted individual rationality can lead to a series of disastrous consequences in the future. The phenomenon of 'inter-temporal tragedy of temporal individual rationality' is becoming increasingly relevant in our age of rapid globalization. Cooperation offers the potential to achieve socially optimal and group-efficient solutions to decision-making problems that involve strategic interactions. The calls for cooperation have not been scarce:

"You look at the large problems that we face — that would be overpopulation, water shortages, global warming, and AIDS, I suppose — all of that needs international cooperation to be solved."

— **Molly Ivins**

"Competition has been shown to be useful up to a certain point and no further, but cooperation, which is the thing we must strive for today, begins where competition leaves off."

— **Franklin D. Roosevelt**

"With closely knitted transnational interests, zero-sum games and conflict confrontations had long been outdated; helping each other on the same boat, cooperation and mutual victory are the need of our time."

— **Xi Jinping**

"The keystone of successful business is cooperation. Friction retards progress."

— **James Cash Penney**

"The hole in the ozone layer is a kind of skywriting. At first, it seemed to spell out our continuing complacency before a witch's brew of deadly perils. But perhaps it really tells of a newfound talent to work together to protect the global environment."

— **Carl Sagan**

"Great discoveries and improvements invariably involve the cooperation of many minds."

— **Alexander Graham Bell**

"Nothing truly valuable can be achieved except by the unselfish cooperation of many individuals."

— **Albert Einstein**

"In the long history of humankind... those who learned to collaborate and improvise most effectively have prevailed."

— **Charles Darwin**

Among the challenges we face regarding cooperation, the most urgent matter is the efficient collaboration in combating environmental degradation. Following several decades of swift technological progress and economic expansion, rising levels of pollution and environmental degradation are surfacing globally. Due to the widespread diffusion of pollutants and the cross-border impacts of ecosystem deterioration, the unilateral actions of a single nation or region frequently prove ineffective. The self-serving optimization by individual countries can lead to catastrophic consequences. Cooperation yields the best promise to resolve the problem.

The most common form of cooperation occurs during the process of globalization, in which numerous joint ventures (JVs) are established. Over the past 30 years, globalization and JVs have contributed to an extraordinary acceleration in global economic growth. JVs are crucial to globalization as they promote collaboration across borders and within domestic markets. By leveraging the intricate structure of the international division of labor through the offshoring of production activities and the transfer of technology, globalization significantly enhances the benefits of trade. This novel production organization is driving transformations that impact the entire value chain across virtually all sectors and types of enterprises. However, globalization and JVs are facing critical challenges, namely, the sustainability of the collaboration, acceptable gain-sharing schemes, and the consistency of the original agreement over time. These unresolved challenges have led to serious consequences, like disruptive terminations of ventures, legal disputes, and bankruptcies. An effective cooperative solution resolving these challenges can greatly increase the gains from trade and build an efficient international business order.

Other major types of cooperation include the following: (i) preferential trade agreements (PTAs), under which countries reduce tariffs on selected goods; (ii) free trade areas (FTAs), in which member countries eliminate tariffs and quotas among themselves; (iii) Pharmaceutical Inspection Co-operation Scheme (PIC/S), which aims to standardize good manufacturing practice regulations globally for regulatory bodies from more than 50 countries; and (iv) scientific and technical cooperation, which promotes sharing research, data, and technology to solve environmental problems through joint climate modeling projects, establishment of international biodiversity databases, and technology transfer programs for clean energy and waste management.

Finally, as pointed out in the 2022 address by United Nations Secretary-General António Guterres, "Our world is in big trouble. Divides are growing deeper. ... The international community is not ready or willing to tackle the big, dramatic challenges of our age. These crises threaten the very future of humanity and the fate of our planet. ... Our world is in peril — and paralyzed. ... No cooperation. No dialogue. No collective problem-solving... But the reality is that we live in a world where the logic of cooperation and dialogue is the only path forward. ... So let's develop common solutions to common problems — grounded in goodwill, trust, and the rights shared by every human being. Let's work as one, a coalition of the world."

Despite the urgent need for cooperation, sustainable collaborative frameworks are rarely observed. Specifically, a successful collaboration must exhibit the following characteristics:

(i) Individual rationality must be maintained, ensuring that the payoff received by any participant in the cooperative arrangement is at least equal to their payoff in a non-cooperative scenario. If individual rationality is not guaranteed, participants may reject the proposed solution and choose to act independently.
(ii) Group optimality guarantees that all potential benefits from cooperation are realized. If group optimality is not achieved, participants may opt to deviate from the agreed solution to capture unexploited benefits.
(iii) Time consistency is essential to ensure that the cooperative agreement remains optimal throughout the duration of the collaboration. Time inconsistency can lead to suboptimal arrangements as the cooperation progresses, potentially resulting in the dissolution of the cooperative scheme.
(iv) A fair gain-sharing mechanism, which is based on performance, contributions, economic rationality, as well as the satisfaction of the perceptions of fairness, is necessary.

A fair gain-sharing principle of cooperative gain is critical to obtaining the agreement of all players to cooperate and ensuring the scheme's sustainability. To fulfill fair gain-sharing, the distribution of the cooperative gains has to reflect the contributions of players. This book presents a fair gain-sharing cooperative solution that satisfies the properties of individual rationality, group optimality, time consistency, and the fair gain-sharing principle that reflects the contributions of the players. The theory and solution techniques are expounded and applied to global environmental management, JVs, situations with random exit, games on networks, and interaction under uncertainty.

The organization of the book is as follows.

Chapter 2 presents the theory and solution techniques of dynamic fair gain-sharing cooperation, and also highlights the best ways to alleviate the harmful effects of individually rational non-cooperative actions. Dynamic cooperation represents one of the most intriguing forms of optimization analysis, and the complexity of the problem leads to great difficulties in deriving satisfactory solutions. Given the asymmetry of the players, some players may capture substantially less gain than their contributions to other players in a cooperative scheme. A proper gain-sharing principle of cooperative gain is critical to obtaining the agreement of all players to cooperate and ensuring the scheme's sustainability. This chapter develops a novel fair gain-sharing solution — fair gain-sharing dynamic cooperative solution — which reflects the contributions of the players to the joint cooperative payoff for cooperative dynamic games. This solution has the following properties: (i) it satisfies individual rationality in that every participant will receive a higher return under cooperation than that under non-cooperation; (ii) it satisfies group optimality in that the total returns under cooperation allocated to the nations coincide with the maximal total returns; (iii) it satisfies the condition of time consistency under which an extension of the solution policy to a time with a later starting stage and a state brought about by prior optimal behaviors would remain optimal; (iv) it reflects the contributions of the participants in the cooperation scheme. Section 1 considers the notion of fair gain-sharing, while Section 2 provides the basic framework for analysis. In Section 3, the idea of fair gain-sharing cooperation is examined. A dynamic fair-sharing cooperative solution is derived in Section 4, while an illustrative example in public capital provision is given in Section 5. Concluding remarks are given in Section 6.

Chapter 3 applies the fair gain-sharing solution to global environmental management. Rapid industrial globalization and economic growth have led to a rapid increase in pollution. Cooperation is the best way to alleviate the harmful effects of individually rational non-cooperative actions in industrial production. Given the asymmetry of the world's nations, a fair gain-sharing scheme is critical to obtaining the agreement of all nations to cooperate and ensuring the scheme's sustainability. This chapter derives a fair gain-sharing solution for global environmental management that satisfies individual rationality, group optimality, time consistency, and the fair gain-sharing property that reflects the contributions of particular nations. Section 1 examines the gain-sharing problems in the context of global environmental management. Section 2 provides the theoretical underpinnings of a remedial solution. Section 3 develops an analytical framework and derives the non-cooperative equilibria and group-optimal outcomes. Section 4 establishes a dynamic fair-sharing cooperative solution for global environmental governance. Section 5 provides an explication of the problems of the Conference of the Parties (COP). Section 6 offers a numerical example, while Section 7 provides the concluding remarks.

Chapter 4 considers dynamic JVs under a fair gain-sharing scheme. JVs have led to unprecedented rapid growth in the global economy over the past three decades, driven by globalization. However, JVs are facing critical challenges, namely, the sustainability of the collaboration, acceptable gain-sharing schemes, and the consistency of the original agreement over time. These unresolved challenges have led to serious consequences, such as the disruptive termination of private-private ventures, international legal disputes involving government-private ventures, and uneventful government-government ventures. This chapter develops a cooperative solution for JVs that entails (i) individual and group rationalities, (ii) a fair gain-sharing scheme, and (iii) the consistency of the agreed-upon agreement. Section 1 discusses JVs and globalization. Section 2 presents a general game framework of JV, the benchmark non-cooperative equilibria, the payoff structure under the JV, and the maximized venture payoff. Section 3 discusses gain-sharing in a dynamic JV. Section 4 presents a dynamic fair-sharing cooperative solution for JVs. Section 5 gives an illustrative example, while Section 6 provides the concluding remarks.

Chapter 5 examines fair gain-sharing under an uncertain cooperation horizon. In many game situations, the terminal time of the game is not known with certainty. Commonly observed examples of horizontal uncertainties include the terms of office of elected authorities, contract renewal, continuation of agreements subjected to periodic negotiations, and the life span of business firms. The demise of firms and the change of political regimes are often not accurately predictable. The presence of fair gain-sharing would be crucial to the formation of a cooperative scheme. This chapter considers fair gain-sharing cooperation under uncertain planning horizons. In Section 1, a dynamic game with random duration is formulated, and a dynamic programming technique for solving inter-temporal problems with a random horizon is developed to serve as the foundation for solving the game problem. In Section 2, the non-cooperative equilibrium and group optimal solution are characterized. Section 3 considers fair gain-sharing cooperation under random exit. The corresponding subgame-consistent solution and payment mechanism are shown in Section 4. An illustration of fair gain-sharing under Random Exit is provided in Section 5, while concluding remarks are given in Section 6.

Chapter 6 studies dynamic fair gain-sharing in games on a network. An analysis of the optimal management of interconnected networks that enhances the payoffs of the network players is presented. An optimality principle in gain-sharing that reflects the contributions of the participating players in a network is derived. A novel dynamic fair gain-sharing solution for interconnected networks that fulfills the requirements of individual rationality, group optimality, time consistency, and the fair-sharing principle. Section 1 provides an overview of networks. Section 2 presents the formulation of an interconnected network, the non-cooperative equilibrium, efficient collaboration, and marginal contributions of players. In Section 3, a dynamic fair gain-sharing collaboration solution that reflects the contributions of the players is presented, and a time-consistent payoff distribution procedure that leads to the achievement of the optimal collaboration solution is derived. An illustrative example of an interconnected electricity network is given in Section 4, while concluding remarks are provided in Section 5.

Chapter 7 considers fair gain-sharing cooperative solutions in interactive situations with uncertain future conditions. The presence of random elements in future structures and developments is prevalent in many practical situations. A gain-sharing scheme reflecting the contributions of the players represents a fair solution to all players amid all the unpredictable future changes. The chapter first formulates the situation in a stochastic dynamic game. A fair gain-sharing time-consistent solution is developed for cooperation purposes. An analytically tractable payoff distribution procedure contingent upon specific random realizations of the state and payoff structure is derived. An extension of the analysis to a system with uncertain future payoff structures and random evolution of the state variables is provided. The game formulation, non-cooperative equilibria, and group optimal conditions in stochastic dynamic games are given in Section 1. Section 2 considers fair gain-sharing cooperation and derives the marginal contributions, the fair gain-sharing weight, and the cooperative solution. An illustration of cooperative resource extraction under uncertainty is presented in Section 3. Section 4 extends the analysis to a system with uncertain future payoff structures and random evolution of the state variables. Section 5 provides the concluding remarks.

It is worth noting that the game structures and derived results across different chapters are distinct, owing to the unique characteristics of the circumstances in each case.

Finally, since human beings live in time and their interactions generally lead to effects over time, it is only a slight exaggeration to maintain that "almost all practically important human interactions can be regarded as dynamic games". A meaningful analysis of cooperation must therefore incorporate the real-life phenomenon of interactions over time. Knowledge of dynamic game theory is preferable, but is not essential for following the analysis and understanding the main results.

Chapter 2

Dynamic Fair Gain-Sharing Cooperation: Theory and Solution Technique

> "Rely on your own strength of body and soul. Take for your star self-reliance, faith, honesty and industry. Don't take too much advice — keep at the helm and steer your own ship, and remember that the great art of commanding is to take a fair share of the work."
>
> — **Noah Porter**

Many real-life interactions are dynamic in nature, and cooperation stands out as one of the most effective methods to mitigate the adverse impacts of individually rational non-cooperative behaviors. Dynamic cooperation is one of the most intriguing forms of optimization analysis, yet the complexity of these problems presents significant challenges in achieving satisfactory solutions. Due to the asymmetry among the participants, certain participants may receive considerably less benefit than their contributions to others within a cooperative framework. An appropriate gain-sharing principle for cooperative gains is essential for securing the consensus of all players to engage in cooperation and ensuring the sustainability of the scheme. In this chapter, we introduce an innovative fair gain-sharing measure — fair gain-sharing dynamic cooperative value — which reflects the contributions of players to the collective cooperative payoff in dynamic cooperative games. With this new scheme, we develop a fair-gain solution that exhibits the following characteristics: (i) it upholds individual rationality, ensuring that each participant receives a greater return through cooperation than through non-cooperation; (ii) it achieves group optimality, where the total returns allocated to the nations under cooperation align with the maximum total returns; (iii) it adheres to the principle of time consistency, whereby extending the solution policy to a later stage and a state resulting from previous optimal actions remains optimal; (iv) it represents the contributions of participants within the cooperative framework. The concept of fair gain-sharing is addressed in Section 1. The foundational framework for analysis is outlined in Section 2. The concept of fair gain-sharing cooperation is explored in Section 3. A dynamic fair-sharing cooperative solution is derived in Section 4. An illustrative example concerning public capital provision is presented in Section 5. Finally, concluding remarks are provided in Section 6.

1. The Notion of Fair Gain-Sharing

An appropriate gain-sharing principle is essential for securing the agreement of all participants to collaborate and guaranteeing the sustainability of the scheme. In this section, we first analyze some existing gain-sharing schemes, then consider the fair gain-sharing scheme, and finally present some examples that need a fair gain-sharing resolution.

1.1. Existing Gain-Sharing Schemes in Dynamic Games

In this section, we examine the three most frequently utilized gain-sharing schemes in dynamic cooperation:

(i) Equal Sharing of Cooperative Gains: The first scheme involves distributing the surplus of cooperative payoffs over the total of non-cooperative payoffs equally among all participating parties. Generally, this principle is favored by smaller participants but tends to be unpopular among larger players with significant size and productive capacity.

(ii) Sharing Gains Proportional to the Relative Size of Non-cooperative Payoffs: The second scheme entails allocating the surplus of cooperative payoffs over the total of non-cooperative payoffs in proportion to the non-cooperative payoffs of the participating parties. Typically, players with greater endowments and payoffs prefer to share gains proportional to non-cooperative payoffs.

A balancing scheme involves distributing the surplus of cooperative payoffs over the total of non-cooperative payoffs based on a weighted average of equal sharing of cooperative gains and proportional (to the relative size of non-cooperative payoffs) gain-sharing. But it is difficult to obtain a commonly acceptable weight.

(iii) The Shapley Value: The third one is the Shapley (1953) value, which is considered to be the most theoretically sophisticated principle to obtain the measure of the participants' contributions in a cooperative scheme. In particular, the Shapley value has to extract the difference between the payoff of any possible coalition (of participants) S and the payoff of the coalition without player i, i.e., $S\backslash i$. The cooperative payoff received by each participant is obtained from the weighted average of these differences in payoffs as

$$\varphi^i(v) = \sum_{S \subseteq N} \frac{(s-1)!(n-s)!}{n!}[v(S) - v(S\backslash i)] \quad \text{for } i \in N.$$

The term $[v(S) - v(S\backslash i)]$ gives the additional value to the payoff of coalition S if player i is kept in coalition S. A weight equaling $\frac{(s-1)!(n-s)!}{n!}$ is attached to $[v(S)-v(S\backslash i)]$ for obtaining a weighted sum for all possible coalitions $S \subseteq N$. However, the weight $\frac{(s-1)!(n-s)!}{n!}$ is the same for all involved coalitions that have s players.

In particular, in the case where the number of players is 2, the Shapley value becomes

$$\begin{aligned}\varphi^i(v) &= \frac{1}{2}[v(\{1,2\}) - v(\{j\})] + \frac{1}{2}[v(\{i\}) - 0)] \\ &= \frac{1}{2}[v(\{1,2\}) - v(\{i\}) - v(\{j\})] + v(\{i\})\end{aligned}$$

for $i, j \in \{1, 2\}$ and $i \neq j$.

Note that $[v(\{1,2\}) - v(\{i\}) - v(\{j\})]$ is the excess of cooperative payoff over the sum of non-cooperative payoffs. Therefore, the excess of cooperative gain over the non-cooperative payoffs is divided equally between player 1 and player 2, irrespective of their relative sizes of contribution.

1.2. Fair Gain-Sharing

Gain-sharing arrangements involve the distribution of the cooperative payoff to the collaborating participants. The decision on the arrangements has to be guided by economic rationality-based arguments and what is perceived as fair. These arrangements play a crucial role in determining whether participants choose to collaborate and ensuring the long-term viability of the schemes. Numerous instances of collaborative scheme failures have been attributed to dissatisfaction with the gain-sharing arrangements. Gain-sharing arrangements are a vital component in establishing sustainable cooperation. Nevertheless, the disparities among participants regarding size, functionality, capability, resources, facilities, and growth potential pose significant challenges in creating gain-sharing arrangements that are acceptable to all involved parties.

The subject of gain-sharing remains a topic of ongoing debate within the academic community. Luo (2009) explored the pursuit of distributive justice in international joint ventures. Mangel and Useem (2000) analyzed the strategic significance of gain-sharing in the labor market. Mohamed *et al.* (2020) introduced a knowledge-based algorithm for gain-sharing aimed at addressing optimization challenges. Di Foggia and Beccarello (2020) investigated the effects of a gain-sharing cost-reflective tariff on waste management expenses under incentive regulation. O'Byrne and Rajgopal (2022) assessed employee value added through a novel measure of gain-sharing between labor and capital. Carlan *et al.* (2022) evaluated cost-effectiveness and gain-sharing scenarios related to the acquisition of a blockchain-based application in the maritime supply chain. Keränen *et al.* (2023) examined gain-sharing within the context of performance-based contracting.

Due to the asymmetry among the players, certain individuals may receive significantly less benefit compared to their contributions to other participants in cooperative arrangements. For example, a participant proficient in producing clean technology will face increased costs in developing such technology for the advantage of others. Conversely, a participant that generates substantial pollution would suffer considerably more economic losses by reducing its environmentally harmful output, which would be advantageous for all other participants. Gain-sharing arrangements pertain to the allocation of benefits to collaborating members in order to achieve a satisfactory outcome based on a performance framework. Within the realm of gain-sharing arrangements, decisions are influenced by arguments rooted in economic rationality and perceptions of fairness. A fair gain-sharing principle for cooperative gains is essential for securing the consensus of all players to collaborate and ensuring the sustainability of the scheme. To achieve fair gain-sharing, the distribution of cooperative gains must reflect the contributions made by the players.

1.3. Need for Fair Gain-Sharing: Examples

While a fair gain-sharing arrangement would support the success of any collaborative initiatives involving asymmetric participants, two specific cooperation scenarios are urgently waiting for a fair gain-sharing scheme to resolve their problems.

(i) Global Environmental Pact: Given the asymmetry of the cooperating nations in an environmental agreement, certain nations may receive significantly less benefits relative to their contributions compared to others. For example, a nation that excels in producing clean technology will face greater costs in developing such technology for the advantage of other nations. At the same time, a nation that generates substantial pollution would suffer considerable economic losses by reducing its environmentally harmful output, which would ultimately benefit all other nations. Over the 30 years of the Conference of the Parties (COP) of the United Nations Framework Convention on Climate Change (UNFCCC), conflicts regarding the allocation of gains, subsidies, contributions, emissions quotas, and the phase-out of technology have always existed. Establishing a fair principle for sharing cooperative gains is essential for securing the agreement of all nations to collaborate and ensuring the sustainability of the initiative.

(ii) Interconnectivity of Electricity Generation: The electricity grid constitutes a complex and extensive system that spans a significant area both locally and internationally. The interconnection of electricity for power networks has the potential to enhance investments in electricity production infrastructure, boost electricity generation, and ensure that member companies have access to dependable electricity supplies by upholding effective emergency response capabilities in the event of disruptions. Furthermore, maintaining these connections serves as a conduit for technology transfer. In summary, the interconnection of electricity providers with mutually beneficial relationships creates substantial avenues for economic benefits and stability in electricity supplies. The instances of electricity network interconnectivity are rapidly increasing on a global scale (see Purvins *et al.* (2016), Meisen & Mohammadi (2010), Zhang *et al.* (2017), Yun & Zhang (2006), Fan *et al.* (2019), and Rafique *et al.* (2018)). The collaboration among interconnected power grid networks offers the most promising opportunity to achieve a collectively optimal solution. Given the asymmetry of the function of the providers, some may capture substantially less gain through interconnection than their contributions to other providers. An optimality principle in gain-sharing that reflects the contributions of the participating providers has to be designed.

2. Basic Framework for Analysis

The best way to characterize interactions of asymmetric agents over time is through a dynamic game framework. In this section, we present the basic framework of discrete-time cooperative dynamic games, the non-cooperative equilibrium, and the group optimal outcome.

2.1. Game Formulation

Consider the T-stage dynamic games with n players. The payoff player i over the stage 1 to stage T is

$$\sum_{k=1}^{T} g_k^i(x_k, u_k^1, u_k^2, \ldots, u_k^n)\delta_1^k + q_{T+1}^i(x_{T+1})\delta_1^{T+1}, \tag{1}$$

where $u_k^i \in U_k^i \subset R^{n_i}$ is the vector of strategies of player $i \in \{1, 2, \ldots, n\} \equiv N$, the vector of state variables of the game at stage

k is $x_k \in X \subset R^m, q^i_{T+1}(x_{T+1})$ is the terminal payoff of player i at stage $T+1$, and δ_1^k is the discount factor from stage 1 to stage k.

The evolution dynamics is governed by the vector difference equation:

$$x_{k+1} = f_k(x_k, u_k^1, u_k^2, \ldots, u_k^n), \quad x_1 = x_1^0. \tag{2}$$

When all the players are acting non-cooperatively, the scenario becomes a dynamic game:

$$\max_{\substack{u_k^i, \\ k \in \{1,2,\ldots,T\}}} \left\{ \sum_{k=1}^{T} g_k^i(x_k, u_k^1, u_k^2, \ldots, u_k^n)\delta_1^k + q_{T+1}^i\left(x_{T+1}\right)\delta_1^{T+1} \right\}, \tag{3}$$

for $i \in N$ subject to dynamics (2).

2.2. Non-cooperative Equilibrium

In this section, we characterize the non-cooperative outcome of the discrete-time dynamic game (1)–(2). A frequently used way to characterize and derive a feedback Nash equilibrium of the game is provided in the following theorem.

Theorem 2.1. *A set of strategies* $\{u_k^{i**} = \varphi_k^i(x)$ *for* $k \in \{1, 2, \ldots, T\}$ *and* $i \in N\}$ *provides a feedback Nash equilibrium solution to the game (2)–(3) if there exist functions* $V^i(k, x)$ *for* $k \in K$ *and* $i \in N$, *such that the following recursive relations are satisfied:*

$$V^i(T+1, x) = q_{T+1}^i(x)\delta_1^{T+1}, \tag{4}$$

$$\begin{aligned} V^i(k, x) = \max_{u_k^i} \Big\{ & g_k^i[x, u_k^{1**}, u_k^{2**}, \ldots, u_k^{i-1**}, u_k^i, u_k^{i+1**}, \ldots, \\ & u_k^{n**}]\delta_1^k + V^i[k+1, f_k[x, u_k^{1**}, u_k^{2**}, \ldots, \\ & u_k^{i-1**}, u_k^i, u_k^{i+1**}, \ldots, u_k^{n**}] \Big\} \end{aligned} \tag{5}$$

for $i \in N$ *and* $k \in \{1, 2, \ldots, T\}$.

Proof. Invoking the discrete-time dynamic programming technique, $V^i(k, x)$ is the maximized payoff of player i for given strategies of the other $n-1$ players $\{u_k^{i**}(x)$ for $j \in N$ and $j \neq i\}$. Hence, a Nash equilibrium appears. □

In particular,

$$V^i(k,x) = \sum_{t=k}^{T} g_t^i(x_t, u_t^{1**}, u_t^{2**}, \ldots, u_t^{n**})\delta_1^t + q_{T+1}^i(x_{T+1})\delta_1^{T+1} \quad (6)$$

is the non-cooperative game equilibrium payoff of player $i \in N$ from stage k to stage $T+1$.

2.3. Group Optimal Outcome

Now, consider the case when the players agree to cooperate and maximize their payoff by solving the problem:

$$\max_{\substack{u_k^j, j \in N \\ k \in \{1,2,\ldots,T\}}} \left\{ \sum_{j=1}^{n} \left[\sum_{k=1}^{T} g_k^j(x_k, u_k^1, u_k^2, \ldots, u_k^n)\delta_1^k + q_{T+1}^j\left(x_{T+1}\right)\delta_1^{T+1} \right] \right\} \quad (7)$$

subject to dynamics (2).

Invoking the discrete-time dynamic programming technique, an optimal solution to the control problems (2) and (7) can be characterized by the following theorem.

Theorem 2.2. *A set of strategies* $\{u_k^{i*} = \psi_k^i(x)$ *for* $k \in \{1, 2, \ldots, T\}$ *and* $i \in N\}$ *provides an optimal solution to the problems* (2) *and* (7) *if there exist functions* $W^N(k,x)$ *for* $k \in \{1, 2, \ldots, T\}$, *such that the following recursive relations are satisfied:*

$$W^N(T+1, x) = \sum_{j=1}^{n} q_{T+1}^j(x)\delta_1^{T+1}, \quad (8)$$

$$W^N(k,x) = \max_{u_k^1, u_k^2, \ldots, u_k^n} \left\{ \sum_{j=1}^{n} g_k^j[x_k, u_k^1, u_k^2, \ldots, u_k^n]\delta_1^k \right.$$
$$\left. + W^N[k+1, f_k(x_k, u_k^1, u_k^2, \ldots, u_k^n)] \right\}$$

$$= \sum_{j=1}^{n} g_k^j[x, u_k^{1*}, u_k^{2*}, \ldots, u_k^{n*}]\delta_1^k$$
$$+ W^N[k+1, f_k(x, u_k^{1*}, u_k^{2*}, \ldots, u_k^{n*})] \qquad (9)$$

for $k \in \{1, 2, \ldots T\}$.

Proof. Equations (8) and (9) satisfy the optimality principle of dynamic programming. Hence, $W^N(k, x)$ is the maximized joint payoff of the players. □

Substituting the optimal control $\{u_k^{i*}$ for $k \in \{1, 2, \ldots, T\}$ and $i \in N\}$ into the state dynamics (2), one can obtain the dynamics of the cooperative trajectory as

$$x_{k+1} = f_k\left(x_k, u_k^{1*}, u_k^{2*}, \ldots, u_k^{n*}\right), \qquad (10)$$

We use $\{x_k^*\}_{k=1}^T$ to denote the solution to (10), which is the optimal cooperative path. Cooperation will stop if any player refuses to continue with the scheme, and they will return to the non-cooperative situation.

In particular,

$$W^N(k, x) = \sum_{t=1}^{n}\left[\sum_{t=k}^{T} g_k^i(x_k, u_k^{1*}, u_k^{2*}, \ldots, u_k^{n*})\delta_1^t + q_{T+1}^i\left(x_{T+1}\right)\delta_1^{T+1}\right] \qquad (11)$$

is the maximized joint payoff from stage k to stage $T+1$.

3. Fair Gain-Sharing Cooperation

Given the asymmetry of the players in a cooperative dynamic game, their contributions to the maximized joint payoff vary. Some players may capture substantially less gain than their contributions to other players in cooperation. A fair distribution scheme has to reflect the contributions of the players. To do this, we first calculate the marginal contribution of a player to the total joint payoff. Then, we provide a fair gain-sharing weight that reflects the contributions of the players in the cooperative scheme.

3.1. Marginal Contributions of Players

To derive a dynamic fair gain-sharing cooperative solution, we first compute the extra value added to the cooperative payoff when an additional player, referred to as player i, is incorporated into the cooperation group. For the sake of notation, we denote the set of players excluding player as $N\backslash i$. The increase in the maximized joint payoff resulting from the inclusion of player is determined by the difference between the maximized joint payoff of all players and the maximized joint payoff in the absence of player i. Given that the contribution to the maximized payoff by player is the difference between the maximized joint payoff of all players and the maximized payoff without player i, we have to solve the game:

$$\max_{\substack{u_k^j, j\in N\backslash i\\ k\in\{1,2,\ldots,T\}}} \left\{ \sum_{j\in N\backslash i} \left[\sum_{k=1}^{T} g_k^j\left(x_k, u_k^1, u_k^2, \ldots, u_k^n\right) \delta_1^k + q_{T+1}^j(x_{T+1})\delta_1^{T+1} \right] \right\}, \tag{12}$$

$$\max_{\substack{u_k^i\\ k\in\{1,2,\ldots,T\}}} \left\{ \sum_{k=1}^{T} g_k^i\left(x_k, u_k^1, u_k^2, \ldots, u_k^n\right) \delta_1^k + q_{T+1}^i(x_{T+1})\delta_1^{T+1} \right\} \tag{13}$$

subject to

$$x_{k+1} = f_k(x_k, u_k^1, u_k^2, \ldots, u_k^n), \quad x_1 = x_1^0. \tag{14}$$

The solution to the game (12)–(14) can be characterized by the following theorem.

Theorem 3.1. *A set of strategies* $\{\hat{u}_k^i = \hat{\varphi}_k^i(x)$ *for* $k \in \{1, 2, \ldots, T\}$ *and* $\{u_k^{(N\backslash i)j} = \varphi_k^{(N\backslash i)j}(x)$ *for* $k \in \{1, 2, \ldots, T\}$ *and* $j \in N\backslash i\}$ *constitutes a set of feedback Nash equilibria of the game* (12)–(14) *if there exists functions* $W^{N\backslash i}(k, x)$ *and* $\hat{V}^i(k, x)$*, which satisfy the following recursive equations:*

$$W^{N\backslash i}(T+1, x) = \sum_{j\in N\backslash i} q_{T+1}^j\left(x_{T+1}\right)\delta_1^{T+1}, \tag{15}$$

$$W^{N\backslash i}(k,x) = \max_{u_k^j, j\in N\backslash i} \Bigg\{ \sum_{j\in N\backslash i} g_k^j(x, u_k^1, u_k^2, \ldots, u_k^{i-1}, \widehat{u}_k^i, u_k^{i+1}, \ldots, u_k^n)\delta_1^k + W^{N\backslash i}[k+1, f_k(x, u_k^1, u_k^2, \ldots, u_k^{i-1}, \widehat{u}_k^i, u_k^{i+1}, \ldots, u_k^n)] \Bigg\}, \tag{16}$$

$$\widehat{V}^i(T+1,x) = q_{T+1}^i(x)\delta_1^{T+1}, \tag{17}$$

$$\widehat{V}^i(k,x) = \max_{u_k^i} \Bigg\{ g_k^i(x, u_k^{(N\backslash i)1}, u_k^{(N\backslash i)2}, \ldots, u_k^{(N\backslash i)i-1}, u_k^i, u_k^{(N\backslash i)i+1}, \ldots, u_k^{(N\backslash i)n})\delta_1^k + \widehat{V}^i[k+1, f_k(u_k^{(N\backslash i)1}, u_k^{(N\backslash i)2}, \ldots, u_k^{(N\backslash i)i-1}, u_k^i, u_k^{(N\backslash i)i+1}, \ldots, u_k^{(N\backslash i)n})] \Bigg\} \tag{18}$$

for $k \in \{1, 2, \ldots, T\}$.

Proof. Conditions (15)–(16) show that $W^{(N\backslash i)}(k,x)$ is the maximized joint payoff of the players in $N\backslash i$ given the game equilibrium strategies $(\hat{u}_k^i)$ of player i. Conditions (17)–(18) show that $\hat{V}^i(k,x)$ is the maximized payoff of the player i given the game equilibrium strategies of the players in $N\backslash i$, that is $\{u_k^{(N\backslash i)j}$ for $j \in N\backslash i\}$. Hence, a Nash equilibrium results. □

The characterization of the game equilibrium of the game in Theorem 2.1 allows one to derive the joint payoff of players if player i is excluded, i.e.,

$$W^{N\backslash i}(k,x) = \sum_{\substack{j=1 \\ j\neq 1}}^{n} \left[\sum_{t=k}^{T} g_t^j(x_t, u_t^{(N\backslash i)1}, u_t^{(N\backslash i)2}, \ldots, u_t^{(N\backslash i)i-1}, \hat{u}_t^i, u_t^{(N\backslash i)i+1}, \ldots, u_t^{(N\backslash i)n})\delta_1^t + q_{T+1}^i\left(x_{T+1}\right)\delta_1^{T+1} \right]. \tag{19}$$

3.2. Fair Gain-Sharing Weight

In a cooperation scheme, the participating players would share the excess of the maximized cooperative joint payoff over the sum of the non-cooperative payoff, i.e.,

$$\left[W^N(k,x) - \sum_{j=1}^{n} V^j(k,x) \right]. \tag{20}$$

A sustainable sharing mechanism has to reflect the players' contributions in the cooperative scheme. To resolve the problem, we develop a gain-sharing weight that reflects the participating players' contributions. Note that the addition to the maximized joint payoff by including player i is the difference between the maximized joint payoff and the maximized joint payoff without player i, i.e.,

$$W^N(k,x) - W^{N\backslash i}(k,x) \quad \text{for } i \in N. \tag{21}$$

To measure player i's share of contribution to the maximized joint payoff when all players' marginal contributions are taken into consideration, we adopt the weight

$$\frac{W^N(k,x) - W^{N\backslash i}(k,x)}{\sum_{\varsigma=1}^{n}[W^N(k,x) - W^{N\backslash \varsigma}(k,x)]} \quad \text{for player } i \in N. \tag{22}$$

The weight (22) reflects the relative size of player i's contribution to the cooperative gain.

4. A Dynamic Fair-Sharing Cooperative Solution

In this section, we first derive a dynamically stable fair gain-sharing solution and then provide a payoff distribution mechanism, leading to the realization of the derived gainsharing solution. In a multistage game, a critical requirement for sustaining cooperation is the concept of time consistency. This concept asserts that the optimality principle agreed upon at the outset must be upheld at all subsequent points in time throughout the game duration along the optimal state trajectory. Time consistency ensures that the motivation to adhere to a commitment during the entire committed period surpasses the motivation to abandon that commitment. It is a vital principle in the development of cooperation strategies. Any cooperation that lacks time consistency is unlikely to be operational for the entire cooperation duration. A cooperation framework that adheres to time consistency is deemed credible and can be relied upon to be executed throughout the committed period. Therefore, the principle of fair gainsharing must be satisfied at every stage of the cooperation period.

4.1. A Dynamically Stable Fair Gain-Sharing Value

Using (20) and (22), we obtain a gain-sharing scheme that allocates the excess of total cooperative payoff over the sum of individual payoff proportionally to the contributions of the players as

$$\frac{W^N(k,x) - W^{N\backslash i}(k,x)}{\sum_{\varsigma=1}^{n}[W^N(k,x) - W^{N\backslash \varsigma}(k,x)]} \times \left[W^N(k,x) - \sum_{j=1}^{n} V^j(k,x)\right] \quad \text{for } i \in N. \tag{23}$$

The imputation to player $i \in N$ is the sum of his fair share of the excess of total cooperative gains plus his non-cooperative gain, i.e.,

$$\frac{W^N(k,x) - W^{N\backslash i}(k,x)}{\sum_{\varsigma=1}^{n}[W^N(k,x) - W^{N\backslash \varsigma}(k,x)]} \times \left[W^N(k,x) - \sum_{j=1}^{n} V^j(k,x)\right] + V^i(k,x). \tag{24}$$

A dynamically stable fair gain-sharing solution which reflects the contributions of the participating players is obtained by maintaining (24) along the cooperative state trajectory $\{x_k^*\}_{k=1}^{T}$.

Hence, a dynamically stable fair gainsharing solution would allot the imputation to player i as

$$\xi^i(k,x_k^*) = \frac{W(k,x_k^*) - W^{N\backslash i}(k,x_k^*)}{\sum_{\varsigma=1}^{n}[W(k,x_k^*) - W^{N\backslash \varsigma}(k,x_k^*)]} \times \left[W(k,x_k^*) - \sum_{j=1}^{n} V^j(k,x_k^*)\right] + V^i(k,x_k^*) \tag{25}$$

for $k \in \{1, 2, \ldots, T\}$ and $i \in N$.

The fair-sharing value in (25) has the following properties:

(i) It satisfies group optimality in that

$$\sum_{i=1}^{n} \xi^i\left(k, x_k^*\right) = \sum_{i=1}^{n} \left\{ \frac{W^N\left(k, x_k^*\right) - W^{N\backslash i}\left(k, x_k^*\right)}{\sum_{\varsigma=1}^{n}\left[W^N\left(k, x_k^*\right) - W^{N\backslash \varsigma}\left(k, x_k^*\right)\right]} \times \left[W^N\left(k, x_k^*\right) - \sum_{j=1}^{n} V^j\left(k, x_k^*\right)\right] + V^i\left(k, x_k^*\right) \right\}$$

$$= \left[W^N\left(k, x_k^*\right) - \sum_{j=1}^{n} V^j\left(k, x_k^*\right)\right] + \sum_{i=1}^{n} V^i\left(k, x_k^*\right)$$

$$= W^N\left(k, x_k^*\right).$$

(ii) It provides a fair measure of the contributions of the companies to the cooperative payoff by sharing the excess of cooperative gain over the sum of the non-cooperative payoffs of all companies according to the proportion of the relative size of the player's contribution to the cooperative gain net of its non-cooperative payoff, i.e.,

$$\frac{W^N\left(k, x_k^*\right) - W^{N\backslash i}\left(k, x_k^*\right)}{\sum_{\varsigma=1}^{n}\left[W^N\left(k, x_k^*\right) - W^{N\backslash \varsigma}\left(k, x_k^*\right)\right]}.$$

(iii) It fulfills time consistency in that the cooperative gain of player i

$$\xi^i\left(k, x_k^*\right) = \frac{W^N\left(k, x_k^*\right) - W^{N\backslash i}\left(k, x_k^*\right)}{\sum_{\varsigma=1}^{n}\left[W^N\left(k, x_k^*\right) - W^{N\backslash \varsigma}\left(k, x_k^*\right)\right]} \times \left[W^N\left(k, x_k^*\right) - \sum_{j=1}^{n} V^j\left(k, x_k^*\right)\right] + V^i\left(k, x_k^*\right)$$

for $i \in N$ is maintained at each stage $k \in \{1, 2, \ldots, T\}$ along the cooperative state trajectory $\{x_k^*\}_{k=1}^{T}$.

(iv) It satisfies individual rationality in that

$$\xi^i\left(k, x_k^*\right) = \frac{W^N\left(k, x_k^*\right) - W^{N\backslash i}\left(k, x_k^*\right)}{\sum_{\varsigma=1}^n \left[W^N\left(k, x_k^*\right) - W^{N\backslash\varsigma}\left(k, x_k^*\right)\right]} \times \left[W\left(k, x_k^*\right) - \sum_{j=1}^n V^j\left(k, x_k^*\right)\right] + V^i\left(k, x_k^*\right) \geq V^i\left(k, x_k^*\right)$$

for $k \in \{1, 2, \ldots, T\}$ and $i \in N$.

4.2. Time-Consistent Payoff Distribution

Crucial to the analysis is the derivation of a payoff distribution procedure (PDP), leading to the realization of the agreed-upon payoffs $\xi^i(k, x_k^*)$ in (25) in all periods $k \in \{1, 2, \ldots, T\}$ throughout the cooperation duration. Invoking Yeung and Petrosyan (2010, 2016), the payment scheme $\beta_k^i(x_k^*)$ provides a subgame-consistent payoff distribution scheme in the sense that the payoff to player i over the periods from k to $T+1$ satisfies the condition

$$\xi^i\left(k, x_k^*\right) = \frac{W\left(k, x_k^*\right) - W^{N\backslash i}\left(k, x_k^*\right)}{\sum_{\varsigma=1}^n \left[W\left(k, x_k^*\right) - W^{N\backslash\varsigma}\left(k, x_k^*\right)\right]} \times \left[W^N\left(k, x_k^*\right) - \sum_{j=1}^n V^j\left(k, x_k^*\right)\right] + V^i\left(k, x_k^*\right) = \beta_k^i\left(x_k^*\right) + \left[\sum_{\zeta=k}^T \beta_\zeta^i\left(x_\zeta^*\right)\delta_1^\zeta + q_{T+1}^i\left(x_{T+1}^*\right)\delta_1^{T+1}\right] \tag{26}$$

for $i \in N$ and $k \in \{1, 2, \ldots, T\}$.

Following Yeung and Petrosyan (2016), the value of $\beta_k^i\left(x_k^*\right)$ can be obtained as follows.

Theorem 4.1. *The agreed-upon imputation* $\xi(k, x_k^*)$ *for* $k \in \{1, 2, \ldots, T\}$ *along the cooperative trajectory* $\{x_k^*\}_{k=1}^T$, *can be realized by a payment*

$$\beta_k^i(x_k^*) = \left(\delta_1^k\right)^{-1}\left[\xi^i(k, x_k^*) - \xi^i\left(k+1, f_k\left(x_k^*, u_k^{1*}, u_k^{2*}, \ldots, u_k^{n*}\right)\right)\right] \tag{27}$$

given to player $i \in N$ *in stage* $k \in \{1, 2, \ldots, T\}$, *which will lead to the realization of the Fair-Sharing Dynamic Cooperative Value in* (25).

Proof. Using (26), one can obtain

$$\begin{aligned}\xi^i&\left(k+1, x_{k+1}^*\right)\\ &= B_{k+1}^i\left(x_{k+1}^*\right)\delta_1^{k+1} + \left\{\sum_{\zeta=k+2}^{T} B_\zeta^i\left(x_\zeta^*\right)\delta_1^\zeta + q_{T+1}^i\left(x_{T+1}^*\right)\delta_1^{T+1}\right\}.\end{aligned} \tag{28}$$

Substituting (27) into (26) yields

$$\xi^i(k, x_k^*) = \beta_k^i(x_k^*)\,\delta_1^k + \xi^i\left(k+1, x_{k+1}^*\right),$$

which can be expressed as

$$\xi^i(k, x_k^*) = \beta_k^i(x_k^*)\,\delta_1^k + \xi^i\left(k+1, f_k\left(x_k^*, u_k^*, u_k^{2*}, \ldots, u_k^{n*}\right)\right). \tag{29}$$

From (20), one can obtain

$$\beta_k^i(x_k^*) = \left(\delta_1^k\right)^{-1}\left[\xi^i(k, x_k^*) - \xi^i\left(k+1, f_k\left(x_k^*, u_k^{1*}, u_k^{2*}, \ldots, u_k^{n*}\right)\right)\right]. \tag{30}$$

□

$\beta_k^i(x_k^* t)$ is the target allocation to player $i \in N$. However, under cooperation, all players would use the cooperative strategies and the payoff that player i will directly receive at stage k along the cooperative trajectory equaling $g_k^i(x_k^*, u_k^{1*}, u_k^{2*}, \ldots, u_k^{n*}))$. Therefore, a side payment

$$\pi_k^i(x_k^*) = \beta_k^i(x_k^*) - g_k^i\left(u_k^{1*}, u_k^{2*}, \ldots, u_k^{n*}\right) \tag{31}$$

has to be given to player $i \in N$ in period for $k \in \{1, 2, \ldots, T\}$ to actually yield the cooperative payoff $\xi^i(k, x_k^*)$.

4.3. Time-Consistent versus Fully Time-Consistent

In addition to individual rationality and group optimality, another crucial factor that affects the sustainability of dynamic cooperation is the condition of time consistency, which requires the extension of the agreed-upon solution policy to remain effective throughout the game horizon. Many of the breakups of cooperative schemes were caused by the failure to maintain the original agreement at later stages. However, even though the original agreement is maintained, there can still be discontentment as the environment changes over time. For instance, an agreed-upon gain-sharing principle is the sharing of gains equally among the players because of their similar conditions at the start of the game. But as cooperation continues, some players develop much faster than others, and they perceive the sharing of the gains proportional to the relative sizes of the non-cooperative payoff to be more preferable.

Remark 4.1. Under the standard condition of time consistency, the initially agreed-upon sharing principle would continue to be effective at any subsequent starting stage, provided that the state is a result of prior optimal behavior. Although the initially agreed-upon sharing principle remains effective, there is no assurance that this principle will continue to be optimal (or acceptable to all participants). For example, if the original sharing principle dictates an equal division of cooperative gains among the participants, over time along the cooperative path, the non-cooperative gains of certain participants may significantly exceed those of others. This creates an incentive to abandon the collaborative venture. Under the fair gain-sharing principle, the originally agreed-upon optimality principle would maintain its optimality (and acceptability to all participants) at any later starting stage, regardless of changes in the environment over time, as long as the state is a result of prior optimal behaviors. We refer to this type of time consistency as fully time-consistent or robust subgame consistency.

The fair gain-sharing solution is not just fair in the sense of reflecting the contributions of the participants; it is also fully time-consistent in the sense that the original agreed-upon optimality principle would remain optimal at any later starting stage.

Remark 4.2. There are three key factors involved in the fair gain-sharing solution:

(1) **Forward-Looking:** The derivation of the non-cooperative equilibrium, the group optimal outcome, and the contributions of the participating players are based on forward-looking dynamic optimization.
(2) **Fair Gain-sharing:** The gain-sharing mechanism involved a fair sharing weight that reflects the contributions of each player.
(3) **Fully Time-consistent:** The gain-sharing principle remains OPTIMAL for the entire duration of the game.

The 3 **F**'s in Remark 4.2 — Forward-Looking, Fair Gain-sharing, and Fully Time-Consistent — are key elements for successful dynamic cooperation.

5. An Illustrative Example in Public Capital Development

Consider a T-stage and n-agent game involving the development of publicly accessible technology/capital. The evolution dynamics of the technology is governed by the difference equation:

$$x_{k+1} = x_k + \sum_{i=1}^{n} a_k^i u_k^i - \gamma_k x_k, \quad x_1 = x_1^0, \tag{32}$$

where $x_k \in R$ is the level of technology in stage $i, u_k^i \in R$ is the investment of agent i in stage $k, a_k^i u_k^i$ is the increment of the technology brought about by agent i's investment u_k^i, and γ_k is the rate of obsolescence of the technology.

Each agent seeks to maximize his payoff:

$$\sum_{k=1}^{T} \left[p_k^i \left(L_k^i x_k\right)^{1/2} - w_k^i L_k^i - c_k^i \left(u_k^i\right)^2 + h_k^i x_k\right] \delta_1^k + \left(h_{T+1}^i x_{T+1} + Q_{T+1}^i\right) \delta_1^{T+1}, \tag{33}$$

where $p_k^i \left(L_k^i x_k\right)^{1/2}$ is the gross revenue from production using the publicly assessable technology and an input $L_k^i, w_k^i L_k^i$ is the cost of the input $L_k^i, c_k^i\left(u_k^i\right)^2$ is the cost of investment, $h_k^i x_k$ denote the benefits from the public technology/capital, $\left(h_{T+1}^i x_{T+1} + Q_{T+1}^i\right)$ is the terminal payoff of agent i in period $T+1$, and δ_1^k is the discount factor from stage 1 to stage k.

The problem of maximizing the nation's objective (33) with target controls subject to dynamics (32) is an n-agent dynamic game.

5.1. Non-cooperative Equilibrium

Invoking Theorem 2.1, a feedback Nash equilibrium solution of the game (32)–(33) can be characterized by the following corollary.

Corollary 5.1. *A set of strategies* $\{u_k^{i**}, L_k^{i**}$ *for* $k \in \{1,2,\ldots,T\}$ *and* $i \in N\}$ *provides a feedback Nash equilibrium solution of the game (32)–(33) if there exist functions* $V^i(k,x)$, *such that the following recursive relations are satisfied:*

$$
\begin{aligned}
V^i(T+1,x) &= \left(h_{T+1}^i x_{T+1} + Q_{T+1}^i\right)\delta_1^{T+1}, \\
V^i(k,x) &= \max_{u_k^t, L_k^i}\left\{\left[p_k^i\left(L_k^i x\right)^{1/2} - w_k^i L_k^i - c_k^i\left(u_k^i\right)^2 + h_k^i x\right]\delta_1^k\right. \\
&\quad \left. + V^i\left(k+1, x + a_k^i u_k^i + \sum_{\substack{j=1 \\ j\neq i}}^{n} a_k^j u_k^{j**} - \gamma_k x\right)\right\} \\
&\quad \text{for } k \in \{1,2,\ldots,T\} \quad \text{and } i \in N.
\end{aligned} \tag{34}
$$

Performing the indicated maximization operator in Corollary 5.1, we can obtain the game equilibrium payoffs and strategies of the agents as follows.

Proposition 5.1. (1) *The game equilibrium payoffs of the agents are*

$$
V^i(k,x) = \left(A_k^i x + C_k^i\right)\delta_1^k,
$$

where

$$A^i_{T+1} = h^i_{T+1}, C^i_{T+1} = Q^i_{T+1},$$

$$A^i_k = \left(\frac{(p^i_k)^2}{4w^i_k} + h^i_k\right) + A^i_{k+1}(1-\gamma_k),$$

$$C^i_k = \left[A^i_{k+1}\left(a^i_k \frac{A^i_{k+1}\delta^{T+1}_k a^i_k}{2c^i_k} + \sum_{\substack{j=1 \\ j\in i}}^{n} a^j_k \frac{A^j_{k+1}\delta^{T+1}_k a^j_k}{2c^j_k}\right) + C^i_{k+1}\right]\delta^{k+1}_k$$
$$\left. - \frac{\left(A^i_{k+1}\delta^{k+1}_k a^i_k\right)^2}{4c^i_k}\right) \tag{35}$$

for $i \in N$ and $k \in \{1, 2, \ldots, T\}$.

(2) *The game equilibrium strategies are*

$$L^{i**}_k = \left(\frac{p^i_k}{2w^i_k}\right)^2 x,$$
$$u^{i**}_k = \frac{A^i_{k+1}\delta^{k+1}_k a^i_k}{2c^i_k} \tag{36}$$

for $i \in N$, and $k \in \{1, 2, \ldots, T\}$.

Proof. See Appendix A. □

5.2. Group Optimal Outcome

Now that the nations agree to cooperate, the joint payoffs of the nations are under collaboration. Invoking Theorem 2.2, an optimal solution under cooperation can be characterized by the following corollary.

Corollary 5.2. *A set of strategies $\{u^{i*}_k, L^{i*}_k$ for $k \in \{1, 2, \ldots, T\}$ and $i \in N\}$ provides an optimal cooperative solution to the game (32)–(33) if there exist functions $W^N(k, x)$ for $k \in \{1, 2, \ldots T\}$, such that* *the following recursive relations are satisfied:*

$$W^N(T+1,x) = \sum_{i=1}^{n}(h_{T+1}^i x_{T+1} + Q_{T+1}^i)\delta_1^{T+1},$$

$$W^N(k,x) = \max_{\substack{u_k^t, L_k^i\\ i\in N}}\left\{\sum_{i=1}^{n}\left(p_k^i\left(L_k^i x\right)^{1/2} - w_k^i L_k^i - c_k^i\left(u_k^i\right)^2 + h_k^i x\right)\delta_k^k\right.$$
$$\left. + W^N\left(k+1, x+\sum_{j=1}^{n} a_k^j u_k^j - \gamma_k x\right)\right\} \tag{37}$$

for $k \in \{1, 2, \ldots, T\}$.

Performing the indicated maximization operator in Corollary 5.2, the maximized joint payoff and the optimal cooperative strategies are obtained in the following proposition.

Proposition 5.2. (1) *The joint payoff of the nations under collaboration is*

$$W^N(k,x) = (A_k x_k + C_k)\,\delta^{k-1}, \tag{38}$$

where

$$A_{T+1} = \sum_{i=1}^{n} h_{T+1}^i, \quad C_{T+1} = \sum_{i=1}^{n} Q_{T+1}^i,$$

$$A_k = \sum_{i=1}^{n}\left(\frac{\left(p_k^i\right)^2}{4w_k^i} + h_k^i\right) + A_{k+1}(1-\gamma_k),$$

$$C_k = \left[A_{k+1}\left(\sum_{j=1}^{n} a_k^j \frac{A_{k+1}\delta_k^{k+1} a_k^j}{2c_k^j}\right) + C_{k+1}\right]\delta_k^{k+1}$$
$$\left. - \sum_{i=1}^{n}\frac{\left(A_{k+1}\delta_k^{k+1} a_k^i\right)^2}{4c_k^i}\right)$$

for $i \in N$ *and* $k \in \{1, 2, \ldots, T\}$.

(2) *The optimal cooperative strategies are*

$$L_k^{i*} = \left(\frac{p_k^i}{2w_k^i}\right)^2 x,$$

$$u_k^{i*} = \frac{A_{k+1}\delta_k^{k+1} a_k^i}{2c_k^i} \tag{39}$$

for $i \in N$, *and* $k \in \{1, 2, \ldots, T\}$.

Proof. See Appendix B. □

Substituting the relevant optimal strategies in (38) into the state dynamics (32), we obtain the optimal cooperative dynamics as

$$x_{k+1} = x_k + \sum_{j=1}^{n} a_k^j \frac{A_{k+1}\delta_k^{k+1} a_k^j}{2c_k^j} - \gamma_k x_k, \quad x_1 = x_1^0. \tag{40}$$

We use $\{x_k^*\}_{k=1}^{T+1}$ to denote the solution to system (40).

Remark 5.1. From Propositions 5.1 and 5.2, one can observe that the marginal valuation of the technology is higher under cooperation than that under noncooperation, i.e., $A_k \geq A_k^i$.

This leads to a higher level of investment for all agents under cooperation and a higher level of technology.

The payoff net of investment cost, i.e., $p_k^i (L_k^i x_k)^{1/2} - w_k^i L_k^i + h_k^i x_k$, is higher for all agents, especially for those with relatively high values of p_k^i and h_k^i.

An agent that produces efficient investment, with low investment cost c_k^i and high technology output a_k^i, would invest more under cooperation, and its contribution to the cooperation scheme could be more than those of the other agents.

In addition, if the agent's gain from the technology is relatively low with low levels of p_k^i and h_k^i, its net payoff, i.e., $p_k^i \left(L_k^i x\right)^{1/2} - w_k^i L_k^i - c_k^i \left(u_k^i\right)^2 + h_k^i x$, may be very small and even negative.

Based on the illustration in Remark 5.1, the participation of the agent depends on a fair gain-sharing arrangement, and a fair gain-sharing cooperative scheme is crucial to the success of cooperation.

5.3. Fair Gain-Sharing Weight

To obtain the fair gain-sharing weight, we first have to obtain $W^{(N\backslash i)}(k, x, \vartheta)$ by solving the game:

$$\max_{\substack{L_k^\ell, u_k^\ell \\ \ell\in N\backslash i \\ k\in\{1,2,\ldots,T\}}} \left\{ \sum_{\ell\in N\backslash i} \left[\sum_{k=1}^{T} \left(p_k^\ell \left(L_k^\ell x_k\right)^{1/2} - w_k^\ell L_k^\ell - c_k^\ell \left(u_k^\ell\right)^2 + h_k^\ell x_k \right) \delta_1^k \right.\right.$$

$$\left.\left. + \left(h_{T+1}^\ell x_{T+1} + Q_{T+1}^\ell \right) \delta_1^{T+1} \right] \right\}, \tag{41}$$

$$\max_{\substack{L_k^i u_k^i \\ k\in\{1,2,\ldots,T\}}} \left\{ \sum_{k=1}^{T} \left(p_k^i \left(L_k^i x_k\right)^{1/2} - w_k^i L_k^i - c_k^i \left(u_k^i\right)^2 + h_k^i x_k \right) \delta_1^k \right.$$

$$\left. + \left(h_{T+1}^i x_{T+1} + Q_{T+1}^i \right) \delta_1^{T+1} \right\} \tag{42}$$

subject to

$$x_{k+1} = x_k + a_k^i u_k^i + \sum_{\ell\in N\backslash i} a_k^\ell u_k^\ell - \gamma_k x_k, x_1 = x_1^0. \tag{43}$$

Invoking Theorem 3.1, a Nash equilibrium of the game (41)–(43) can be characterized by the following corollary.

Corollary 5.3. *A set of strategies* $\{\hat{u}_k^i$ *for* $k \in \{1, 2, \ldots, T\}$ *and* $\{u_k^{(N\backslash i)j}$ *for* $k \in \{1, 2, \ldots, T\}$ *and* $j \in N\backslash i\}$ *constitutes a set of feedback Nash equilibria of the game (41)–(43) if there exist functions* $W^{N\backslash i}(k, x)$ *and* $\hat{V}^i(k, x)$*, which satisfy the following recursive equations:*

$$W^{N\backslash i}(T+1, x) = \sum_{\ell\in N\backslash i} \left(h_{T+1}^\ell x_{T+1} + Q_{T+1}^\ell \right) \delta_1^{T+1},$$

$$\widehat{V}^i(T+1, x) = \left(h_{T+1}^i x_{T+1} + Q_{T+1}^i \right) \delta_1^{T+1}, \tag{44}$$

$$W^{N\backslash i}(k,x) = \max_{\substack{L_k^\ell, u_k^\ell \\ \ell \in N\backslash i}} \left\{ \sum_{\ell \in N\backslash i} \left(p_k^\ell \left(L_k^\ell x_k \right)^{1/2} \right. \right.$$

$$\left. - w_k^\ell L_k^\ell - c_k^\ell \left(u_k^\ell \right)^2 + h_k^\ell x_k \right) \delta_1^k$$

$$\left. + W^{N\backslash i} \left(k+1, x + a_k^i \widehat{u}_k^i + \sum_{\ell \in N\backslash i} a_k^\ell u_k^\ell - \gamma_k x_k \right) \right\}$$

for $k \in \{1, 2, \ldots, T\}$;

$$\hat{V}^i(k,x) = \max_{\&L_k^i, u_k^t} \left\{ \left[p_k^i \left(L_k^i x \right)^{\frac{1}{2}} - w_k^i L_k^i - c_k^i \left(u_k^i \right)^2 + h_k^i x \right] \delta_1^k \right.$$

$$\left. + \hat{V}^i \left(k+1, x + a_k^i u_k^i + \sum_{\substack{j=1 \\ j \neq i}}^{n} a_k^j u_k^{(N\backslash i)j} - \gamma_k x \right) \right\} \tag{45}$$

for $k \in \{1, 2, \ldots, T\}$.

Performing the indicated maximization operators in Corollary 5.3, we obtain the game equilibrium payoffs and strategies in the following proposition.

Proposition 5.3. (1a) *The game equilibrium joint payoff of the agents in* $N\backslash i$ *at stage* k *is*

$$W^{N\backslash i}(k,x) = \left(A_k^{N\backslash i} x_k + C_k^{N\backslash i} \right) \delta^{k-1}, \tag{46}$$

where

$$A_{T+1}^{N\backslash i} = \sum_{\ell \in N\backslash i} h_{T+1}^\ell, \quad C_{T+1}^{N\backslash i} = \sum_{\ell \in N\backslash i} Q_{T+1}^\ell,$$

$$A_k^{N\backslash i} = \sum_{\ell \in N\backslash i} \left(\frac{(p_k^\ell)^2}{4 w_k^\wp} + h_k^\ell \right) + A_{k+1}^{N\backslash i}(1 - \gamma_k),$$

$$C_k^{N\backslash i} = \left[A_{k+1} \left(a_k^i \frac{\hat{A}_{k+1}^i \delta_k^{k+1} a_k^i}{2c_k^i} + \sum_{j \in N\backslash i} a_k^j \frac{A_{k+1}^{N\backslash i} \delta_k^{k+1} a_k^j}{2c_k^j} - \sum_{j \in N\backslash i} \frac{\left(A_{k+1}^{N\backslash i} \delta_k^{k+1} a_k^j \right)^2}{4c_k^j} \right) + C_{k+1}^{N\backslash i} \right] \delta_k^{k+1}$$

for $k \in \{1, 2, \ldots, T\}$.

(1b) *The game equilibrium payoff of agent* i *in stage* k *is* $\left(\hat{A}_k^i x_k + \hat{C}_k^i \right) \delta_1^k$, *where*

$$\hat{A}_{T+1}^i = h_{T+1}^i, \quad \hat{C}_{T+1}^i = Q_{T+1}^i,$$

$$\hat{A}_k^i = \left(\frac{\left(p_k^i\right)^2}{4w_k^i} + h_k^i \right) + \hat{A}_{k+1}^i \left(1 - \gamma_k\right),$$

$$\hat{C}_k^i = \left[\hat{A}_{k+1}^i \left(a_k^i \frac{\hat{A}_{k+1}^i \delta_k^{k+1} a_k^i}{2c_k^i} + \sum_{j \in N\backslash i} a_k^j \frac{A_{k+1}^{(N\backslash i)j} \delta_k^{k+1} a_k^j}{2c_k^j} \right) + \hat{C}_{k+1}^i \right] \delta_k^{k+1} - \frac{\left(\hat{A}_{k+1}^i \delta_k^{k+1} a_k^i \right)^2}{4c_k^i}. \tag{47}$$

(2) *The game equilibrium strategies are*

$$L_k^{(N\backslash i)\ell} = \left(\frac{p_k^\ell}{2w_k^\ell} \right)^2 x,$$

$$u_k^{(N\backslash i)\ell} = \frac{A_{k+1}^{N\backslash i} \delta_k^{k+1} a_k^\ell}{2c_k^\ell} \quad \text{for } \ell \in N\backslash i,$$

$$\hat{L}_k^i = \left(\frac{p_k^i}{2w_k^i} \right)^2 x,$$

$$\hat{u}_k^i = \frac{\hat{A}_{k+1} \delta_k^{k+1} a_k^i}{2c_k^i} \tag{48}$$

for $k \in \{1, 2, \ldots, T\}$.

Proof. See Appendix C. □

A fair gain-sharing weight can be obtained as

$$\frac{W^N(k,x_k) - W^{N\backslash i}(k,x_k)}{\sum_{\varsigma=1}^{n}\left[W^N(k,x_k) - W^{N\backslash \varsigma}(k,x_k)\right]} = \frac{A_k x_k + C_k - \left(A_k^{N\backslash i} x_k + C_k^{N\backslash i}\right)}{\sum_{\varsigma=1}^{n}\left[A_k x_k + C_k - \left(A_k^{N\backslash \varsigma} x_k + C_k^{N\backslash \varsigma}\right)\right]} \tag{49}$$

for $i \in N$.

5.4. A Dynamically Stable Fair Gain-Sharing Solution

Using (25), a dynamically stable fair gain-sharing value can be obtained as

$$\xi^i(k,x_k) = \frac{W^N(k,x_k) - W^{N\backslash i}(k,x_k)}{\sum_{\zeta=1}^{n}\left[W^N(k,x_k) - W^{N\backslash \varsigma}(k,x_k)\right]} \times \left[W^N(k,x_k) - \sum_{j=1}^{n} V^j(k,x_k)\right] + V^i(k,x_k) \tag{50}$$

for $i \in N$ along the cooperative state path $\{x_k^*\}_{k=1}^T$.

6. Concluding Remarks

Gain-sharing has historically posed a challenge for participants engaged in a dynamic cooperative scheme. This chapter invokes the economic rationality that the sharing of gains has to be based on the contributions of the participants. A mechanism is developed to derive a weight that quantifies the relative contributions of the participants. Utilizing this weight, we propose an innovative fair gain-sharing solution for cooperative dynamic games. Specifically, this fair gain-sharing solution exhibits the characteristics of (i) individual rationality, (ii) group optimality, (iii) time consistency, and (iv) adherence to the fair gain-sharing principle. A fair-sharing solution that reflects the contributions of players in cooperative dynamic network games was formulated by Yeung et al. (2024, 2025a), and a fair gain-sharing solution for dynamic games was given in Yeung and Petrosyan (2025).

In the case where the game involves only one stage of operation, the game structure involves the payoffs of player $i \in N$ being $g_1^i(x_1, u_1^1, u_1^2, \ldots, u_1^n)$.

The non-cooperative game equilibrium payoff, i.e., $V^i(1, x_1)$, can be obtained by solving the static game:

$$\max_{u_1^i} \left\{ g_1^i \left(x_1, u_1^1, u_1^2, \ldots, u_1^n \right) \right\} \tag{51}$$

for $i \in N$.

The maximized joint payoffs under cooperation, i.e., $W^N(1, x_1)$, can be obtained by solving the optimization problem in (51).

The maximized joint payoffs of coalition $N\backslash i$, i.e., $W^{N\backslash i}(1, x_1)$, can be obtained by solving the game:

$$\max_{u_1^j, j \in N\backslash i} \left\{ \sum_{j \in N\backslash i} g_1^j \left(x_1, u_1^1, u_1^2, \ldots, u_1^n \right) \right\} \tag{52}$$

and

$$\max_{u_1^i} \left\{ g_1^i \left(x_1, u_1^1, u_1^2, \ldots, u_1^n \right) \right\}. \tag{53}$$

We can obtain the payoff allotted to player $i \in N$ as

$$\begin{aligned} \xi^i(1, x_1) = {} & \frac{W(1, x_1) - W^{N\backslash i}(1, x_1)}{\sum_{\varsigma=1}^{n} \left[W(1, x_1^*) - W^{N\backslash \varsigma}(1, x_1) \right]} \\ & \times \left[W(1, x_1) - \sum_{j=1}^{n} V^j(1, x_1) \right] + V^i(1, x_1). \end{aligned} \tag{54}$$

Using the notation in the Shapley value, (54) can be expressed as

$$\varphi^i(v) = \frac{v(N) - v(N\backslash i)}{\sum_{j=1}^{n} [v(N) - v(N\backslash j)]} \left[v(N) - \sum_{j=1}^{n} v(j) \right] + v(i). \tag{55}$$

A Comparative Remark on the Shapley value and the fair gain-sharing value is given in the following.

Remark 6.1.

(i) **Computation Time:** To compute the Shapley value, one has to consider the payoffs of 2^n coalitions in a n-person game. For the fair gain-sharing value, a total of $2n+1$ payoffs are involved.
(ii) **Avoidance of Arbitrary Equal Sharing:** If the number of players is 2, the Shapley value implies equal sharing of the cooperative payoff. For the fair gain-sharing value, the weight always leads to a value that reflects the contributions of the participants.
(iii) **Contributions Measured:** The Shapley value considers the additional value of player i to coalition S, that is $[v(S)-v(S\backslash i)]$, with a weight $\frac{(s-1)!(n-s)!}{n!}$. The fair gain-sharing value considers the additional value contributed by player i to the joint payoff $[v(N)-\sum_{j=1}^{n} v(\{j\})]$ with a contribution-guided weight.
(iv) **Underlying Optimization Foundation:** In considering the payoffs $v(S)$ and $v(S\backslash i)$, the Shapley value has no definite specification of the behaviors of the other players $j \in N\backslash S$. The fair gain-sharing value explicitly specifies the optimization process for the relevant terms.

Appendices

Appendix A. Proof of Proposition 5.1

Using the first equation in Corollary 5.1, we obtain

$$\begin{aligned} V^i(T+1,x) &= \left(h^i_{T+1}x_{T+1} + Q^i_{T+1}\right)\delta_1^{T+1} \\ &\equiv \left(A^i_{T+1}x_{T+1} + C^i_{T+1}\right)\delta_1^{T+1}. \end{aligned} \tag{A.1}$$

Invoking the technique of backward induction, we consider first the last stage of operation, i.e., stage T. Performing the indicated maximization operator in Corollary 5.1, we can obtain the game equilibrium strategies in stage T as

$$\begin{aligned} L_T^i &= \left(\frac{p_T^i}{2w_T^i}\right)^2 x, \\ u_T^i &= \frac{A_{T+1}^i \delta_T^{T+1} a_T^i}{2c_T^i} \end{aligned} \tag{A.2}$$

for $i \in N$.

Substituting the game equilibrium target controls in (A.2) into the stage T equation in Corollary 5.1, we obtain

$$\begin{aligned} V^i(T,x) = &\left(\frac{(p_T^i)^2}{4w_T^i}x - \frac{\left(A_{T+1}^i \delta_T^{T+1} a_T^i\right)^2}{4c_T^i} + h_T^i x\right)\delta_1^T \\ &+ \left[A_{T+1}^i\left(x + a_T^i \frac{A_{T+1}^i \delta_T^{T+1} a_T^i}{2c_T^i}\right.\right. \\ &\left.\left. + \sum_{\substack{j=1 \\ j\in i}}^{n} a_T^j \frac{A_{T+1}^j \delta_T^{T+1} a_T^j}{2c_T^j} - \gamma_T x\right) + C_{T+1}^i\right]\delta_1^{T+1} \end{aligned} \tag{A.3}$$

for $i \in N$.

The RHS of the equation system (A.3) is a linear function of x, which can be expressed as $\left(A_T^i x_T + C_T^i\right)\delta_1^T$, where

$$\begin{aligned} A_T^i &= \left(\frac{(p_T^i)^2}{4w_T^i} + h_T^i\right) + A_{T+1}^i(1-\gamma_T), \\ C_T^i &= \left[A_{T+1}^i\left(a_T^i \frac{A_{T+1}^i \delta_T^{T+1} a_T^i}{2c_T^i} + \sum_{j\in 1}^{n} a_T^j \frac{A_{T+1}^j \delta_T^{T+1} a_T^j}{2c_T^j}\right) + C_{T+1}^i\right]\delta_T^{T+1} \\ &\left. - \frac{\left(A_{T+1}^i \delta_T^{T+1} a_T^i\right)^2}{4c_T^i}\right). \end{aligned} \tag{A.4}$$

Following the analysis in (A.1)–(A.4), we can obtain the game equilibrium strategies in stage $k \in \{T-1, T-2, \ldots, 1\}$ as

$$L_k^i = \left(\frac{p_k^i}{2w_k^i}\right)^2 x,$$
$$u_k^i = \frac{A_{k+1}^i \delta_k^{k+1} a_k^i}{2c_k^i} \tag{A.5}$$

for $i \in N$.

The game equilibrium payoff of the agents in stage k can be obtained as $\left(A_k^i x_k + C_k^i\right) \delta_1^k$, where

$$A_k^i = \left(\frac{\left(p_k^i\right)^2}{4w_k^i} + h_k^i\right) + A_{k+1}^i \left(1 - \gamma_k\right),$$
$$C_k^i = \left[A_{k+1}^i \left(a_k^i \frac{A_{k+1}^i \delta_k^{k+1} a_k^i}{2c_k^i} + \sum_{\substack{j=1 \\ j \in i}}^{n} a_k^j \frac{A_{k+1}^j \delta_k^{k+1} a_k^j}{2c_k^j}\right) + C_{k+1}^i\right] \delta_k^{k+1}$$
$$- \frac{\left(A_{k+1}^i \delta_k^{k+1} a_k^i\right)^2}{4c_k^i} \tag{A.6}$$

for $i \in N$.

Appendix B. Proof of Proposition 5.2

Using the first equation in Corollary 5.2, we obtain

$$\begin{aligned} W^N(T+1, x) &= \sum_{i=1}^{n} \left(h_{T+1}^i x_{T+1} + Q_{T+1}^i\right) \delta_1^{T+1} \\ &= \left(\sum_{i=1}^{n} h_{T+1}^i x_{T+1} + \sum_{i=1}^{n} Q_{T+1}^i\right) \delta_1^{T+1} \\ &\equiv \left(A_{T+1} x_{T+1} + C_{T+1}\right) \delta_1^{T+1}. \end{aligned} \tag{B.1}$$

Invoking the technique of backward induction, we consider first the last stage of operation, i.e., stage T. Performing the indicated maximization operator in Corollary 5.2, we can obtain the optimal cooperative strategies in stage T as

$$L_T^i = \left(\frac{p_T^i}{2w_T^i}\right)^2 x,$$
$$u_T^i = \frac{A_{T+1}\delta_T^{T+1} a_T^i}{2c_T^i} \tag{B.2}$$

for $i \in N$.

Substituting the optimal cooperative strategies in (B.2) into the stage T equation of Corollary 5.2, we obtain

$$\begin{aligned} W^N(T,x) = {} & \sum_{i=1}^{n}\left(\frac{(p_T^i)^2}{4w_T^i}x - \frac{\left(A_{T+1}\delta_T^{T+1}a_T^i\right)^2}{4c_T^i} + h_T^i x\right)\delta_1^T \\ & + \left[A_{T+1}\left(x + \sum_{j=1}^{n} a_T^j \frac{A_{T+1}\delta_T^{T+1}a_T^j}{2c_T^j} - \gamma_T x\right)\right. \\ & \left. + C_{T+1}^T\right]\delta_1^{T+1}. \end{aligned} \tag{B.3}$$

The RHS of equation (B.3) is a linear function of x, which can be expressed as $(A_T x_T + C_T)\delta_1^T$, where

$$\begin{aligned} A_T &= \sum_{i=1}^{n}\left(\frac{(p_T^i)^2}{4w_T^i} + h_T^i\right) + A_{T+1}(1-\gamma_T), \\ C_T &= \left[A_{T+1}\left(\sum_{j=1}^{n} a_T^j \frac{A_{T+1}\delta_T^{T+1}a_T^j}{2c_T^j}\right) + C_{T+1}\right]\delta_T^{T+1} \\ & \quad - \sum_{i=1}^{n}\frac{\left(A_{T+1}\delta_T^{T+1}a_T^i\right)^2}{4c_T^i}. \end{aligned} \tag{B.4}$$

Following the analysis in (B.1)–(B.4), we can obtain the optimal cooperative strategies in stage $k \in \{T-1, T-2, \ldots, 1\}$ as

$$
\begin{aligned}
L_k^i &= \left(\frac{p_k^i}{2w_k^i}\right)^2 x, \\
u_k^i &= \frac{A_{k+1}\delta_k^{k+1} a_k^i}{2c_k^i}
\end{aligned}
\tag{B.5}
$$

for $i \in N$.

The maximized joint payoff of the agents in stage k can be obtained as $(A_k x_k + C_k)\,\delta_1^k$, where

$$
\begin{aligned}
A_k &= \sum_{i=1}^{n}\left(\frac{\left(p_k^i\right)^2}{4w_k^i} + h_k^i\right) + A_{k+1}\,(1-\gamma_k), \\
C_k &= \left[A_{k+1}\left(\sum_{j=1}^{n} a_k^j \frac{A_{k+1}\delta_k^{k+1} a_k^j}{2c_k^j}\right) + C_{k+1}\right]\delta_k^{k+1} \\
&\quad - \sum_{i=1}^{n}\frac{\left(A_{k+1}\delta_k^{k+1} a_k^i\right)^2}{4c_k^i}.
\end{aligned}
\tag{B.6}
$$

Appendix C. Proof of Proposition 5.3

Using equation (44) in Corollary 5.3, we obtain

$$
\begin{aligned}
W^{N\backslash i}(T+1, x) &= \sum_{\ell \in N\backslash i}\left(h_{T+1}^{\ell} x_{T+1} + Q_{T+1}^{\ell}\right)\delta_1^{T+1} \\
&= \left(\sum_{\ell \in N\backslash i} h_{T+1}^{\ell} x_{T+1} + \sum_{\ell \in N\backslash i} Q_{T+1}^{\ell}\right)\delta_1^{T+1} \\
&\equiv \left(A_{T+1}^{N\backslash i} x_{T+1} + C_{T+1}^{N\backslash i}\right)\delta_1^{T+1},
\end{aligned}
\tag{C.1}
$$

$$
\begin{aligned}
\hat{V}^i(T+1, x) &= \left(h_{T+1}^i x_{T+1} + Q_{T+1}^i\right)\delta_1^{T+1} \\
&\equiv \left(A_{T+1}^i x_{T+1} + C_{T+1}^i\right)\delta_1^{T+1}.
\end{aligned}
\tag{C.2}
$$

Invoking the technique of backward induction, we consider first the last stage of operation, i.e., stage T. Performing the indicated maximization operator in Corollary 5.2, we can obtain the optimal cooperative strategies in stage T as

$$L_T^\ell = \left(\frac{p_T^\ell}{2w_T^i}\right)^2 x,$$

$$u_T^\ell = \frac{A_{T+1}^{N\backslash i}\delta_T^{T+1}a_T^\ell}{2c_T^\ell}$$

for $\ell \in N\backslash i$;

$$L_T^i = \left(\frac{p_T^i}{2w_T^i}\right)^2 x,$$

$$u_T^i = \frac{A_{T+1}^i\delta_T^{T+1}a_T^i}{2c_T^i} \tag{C.3}$$

for $i \in N$.

Substituting the game equilibrium strategies in (C.2) into the stage T equations of Corollary 5.3, we obtain

$$\begin{aligned} W^{N\backslash i}(T,x) = {} & \sum_{\ell\in N\backslash i}\left(\frac{(p_T^\ell)^2}{4w_T^\ell}x - \frac{\left(A_{T+1}^{N\backslash i}\delta_T^{T+1}a_T^\ell\right)^2}{4c_T^\ell} + h_T^\ell x\right)\delta_1^T \\ & + \left[A_{T+1}^{N\backslash i}\left(x + a_T^i\frac{\hat{A}_{T+1}^i\delta_T^{T+1}a_T^i}{2c_T^i}\right.\right. \\ & \left.\left. + \sum_{j\in N\backslash i} a_T^j\frac{A_{T+1}^{N\backslash i}\delta_T^{T+1}a_T^j}{2c_T^j} - \gamma_T x\right) + C_{T+1}^{N\backslash i}\right]\delta_1^{T+1}, \end{aligned} \tag{C.4}$$

$$\hat{V}^i(T,x) = \left(\frac{\left(p_T^i\right)^2}{4w_T^i} x - \frac{\left(A_{T+1}^i \delta_T^{T+1} a_T^i\right)^2}{4c_T^i} + h_T^i x \right) \delta_1^T$$
$$+ \left[\hat{A}_{T+1}^i \left(x + a_T^i \frac{\hat{A}_{T+1}^i \delta_T^{T+1} a_T^i}{2c_T^i} + \sum_{j \in N \backslash i} a_T^j \right.\right.$$
$$\left.\left. \times \frac{A_{T+1}^{(N\backslash i)j} \delta_T^{T+1} a_T^j}{2c_T^j} - \gamma_T x \right) + \hat{C}_{T+1}^i \right] \delta_1^{T+1}. \quad \text{(C.5)}$$

The RHS of equations (C.4)–(C.5) are linear functions of x which can be expressed as $\left(A_T^{N\backslash i} x_T + C_T^{N\backslash i}\right) \delta_1^T$, where

$$A_T^{N\backslash i} = \sum_{\ell \in N\backslash i} \left(\frac{\left(p_T^\ell\right)^2}{4w_T^\ell} + h_T^\ell \right) + A_{T+1}^{N\backslash i} (1 - \gamma_T),$$

$$C_T^{N\backslash i} = \left[A_T^{N\backslash i} \left(a_T^i \frac{\hat{A}_{T+1}^i \delta_T^{T+1} a_T^i}{2c_T^i} + \sum_{j \in N\backslash i} a_T^j \frac{A_{T+1}^{N\backslash i} \delta_T^{T+1} a_T^j}{2c_T^j} \right) \right.$$

$$\left. + C_{T+1}^{N\backslash i} \right] \delta_T^{T+1} - \sum_{\ell \in N\backslash i} \frac{\left(A_{T+1}^{N\backslash i} \delta_T^{T+1} a_T^\ell\right)^2}{4c_T^\ell},$$

$(\hat{A}_T^i x_T + \hat{C}_T^i)\delta_1^T$, where

$$\hat{A}_T^i = \left(\frac{\left(p_T^i\right)^2}{4w_T^i} + h_T^i \right) + \hat{A}_{T+1}^i (1 - \gamma_T),$$

$$\hat{C}_T^i = \left[\hat{A}_{T+1}^i \left(a_T^i \frac{\hat{A}_{T+1}^i \delta_T^{T+1} a_T^i}{2c_T^i} + \sum_{j \in N\backslash i} a_T^j \frac{A_{T+1}^{(N\backslash i)j} \delta_T^{T+1} a_T^j}{2c_T^j} \right) \right.$$

$$\left. + \hat{C}_{T+1}^i \right] \delta_T^{T+1} - \frac{\left(\hat{A}_{T+1}^i \delta_T^{T+1} a_T^i\right)^2}{4c_T^i} \Bigg). \quad \text{(C.6)}$$

Following the analysis in (C.1)–(C.4), we can obtain the game equilibrium strategies in stage $k \in \{T-1, T-2, \ldots, 1\}$ as

$$L_k^{(N\backslash i)\ell} = \left(\frac{p_k^\ell}{2w_k^\ell}\right)^2 x,$$
$$u_k^\ell = \frac{A_{k+1}^{N\backslash i}\delta_k^{k+1}a_k^\ell}{2c_k^\ell}$$

for $\ell \in N\backslash i$,

$$L_k^i = \left(\frac{p_k^i}{2w_k^i}\right)^2 x,$$
$$u_k^i = \frac{\hat{A}_{k+1}\delta_k^{k+1}a_k^i}{2c_k^i}. \tag{C.7}$$

The maximized joint payoff of the agents in $N\backslash i$ stage k can be obtained as $\left(A_k^{N\backslash i}x_k + C_k^{N\backslash i}\right)\delta_1^k$, where

$$A_k^{N\backslash i} = \sum_{\ell\in N\backslash i}\left(\frac{\left(p_k^\ell\right)^2}{4w_k^\ell} + h_k^\ell\right) + A_{k+1}\left(1-\gamma_k\right),$$
$$C_k^{N\backslash i} = \left[A_{k+1}^{N\backslash i}\left(a_k^i\frac{\hat{A}_{k+1}^i\delta_k^{k+1}a_k^i}{2c_k^i} + \sum_{j\in N\backslash i}a_k^j\frac{A_{k+1}^{N\backslash i}\delta_k^{k+1}a_k^j}{2c_k^j}\right)\right.$$
$$\left. + C_{k+1}^{N\backslash i}\right]\delta_k^{k+1} - \sum_{\ell\in N\backslash i}\frac{\left(A_{k+1}^{N\backslash i}\delta_k^{k+1}a_k^\ell\right)^2}{4c_k^\ell}. \tag{C.8}$$

The game equilibrium payoff of agent i in stage k can be obtained as $\left(\hat{A}_k^i x_k + \hat{C}_k^i\right)\delta_1^k$, where

$$\hat{A}_k^i = \left(\frac{\left(p_k^i\right)^2}{4w_k^i} + h_k^i\right) + \hat{A}_{k+1}^i\left(1-\gamma_k\right),$$

$$\hat{C}_k^i = \left[\hat{A}_{k+1}^i \left(a_k^i \frac{\hat{A}_{k+1}^i \delta_k^{k+1} a_k^i}{2c_k^i} + \sum_{j \in N \backslash i} a_k^j \frac{A_{k+1}^{(N \backslash i)j} \delta_k^{k+1} a_k^j}{2c_k^j}\right) + \hat{C}_{k+1}^i\right] \delta_k^{k+1} - \frac{\left(\hat{A}_{k+1}^i \delta_k^{k+1} a_k^i\right)^2}{4c_k^i} \quad \text{for } i \in N. \tag{C.9}$$

References

Carlan, V., Sys, C., & Vanelslander, T. (2022). Cost-effectiveness and gain-sharing scenarios for purchasing a blockchain-based application in the maritime supply chain. *European Transport Research Review*, 14, 21. https://doi.org/10.1186/s12544-022-00545-2.

Di Foggia, G., & Beccarello, M. (2020). The impact of a gain-sharing cost-reflective tariff on waste management cost under incentive regulation: The Italian case. *Journal of Environmental Management*, 265, 110526–110526. https://doi.org/10.1016/j.jenvman.2020.110526.

Fan, J., Wang, X., Huang, Q., Zhang, X., Li, Y., & Zeng, P. (2019). Power grid interconnection with HVDC link in Northeast Asia considering complementarity of renewable energy and time zone difference. *The Journal of Engineering*, 2019(16), 1625–1629.

Keränen, J., Kienzler, M., Salonen, A., Terho, H., & Totzek, D. (2023). Gain-sharing in performance-based contracting: How risk and fairness drive business customers' willingness-to-switch to a gain-sharing arrangement. *Industrial Marketing Management*, 115, 172–184. https://doi.org/10.1016/j.indmarman.2023.09.013.

Luo, Y. (2009). From gain-sharing to gain-generation: The quest for distributive justice in international joint ventures. *Journal of International Management*, 15(4), 343–356. https://doi.org/10.1016/j.intman.2008.12.006

Mangel, R., & Useem, M. (2000). The strategic role of gainsharing. *Journal of Labor Research*, 21, 327–343. https://doi.org/10.1007/s12122-000-1051-y.

Meisen, P., & Mohammadi, C. (2010). *Cross-Border Interconnections on Every Continent.* Global Energy Network Institute, San Diego, CA, USA. 33.

Mohamed, A. W., Hadi, A. A., & Mohamed, A. K. (2020). Gaining-sharing knowledge based algorithm for solving optimization problems: A novel nature-inspired algorithm. *International Journal of Machine Learning and Cybernetics*, 11(7), 1501–1529. https://doi.org/10.1007/s13042-019-01053-x.

O'Byrne, S. F., & Rajgopal, S. (2022). Employee value added: a new measure of gain-sharing between labor and capital. *Journal of Applied Corporate Finance*, 34(2), 30–44. https://doi.org/10.1111/jacf.12503.

Purvins, A., Fulli, G., Covrig, C. F., Chaouachi, A., Bompard, E., Carpaneto, E., Huang, T., Pi, R. J., Mutule, A., Oleinikova, I., & Obushevs, A. (2016). The Baltic Power System between East and West Interconnections. *Publications Office of the European Union: Luxembourg.*

Rafique, S. F., Shen, P., Wang, Z., Rafique, R., Iqbal, T., Ijaz, S., & Javaid, U. (2018). Global power grid interconnection for sustainable growth: Concept, project and research direction. *IET Generation, Transmission & Distribution*, 12(13), 3114–3123.

Shapley, L.S. (1953). A Value for N-person Games, In H. Kuhn and A. Tucker (eds.), *Contributions to the Theory of Games.* Princeton University Press, Princeton, pp. 307–317.

Yeung, D. W. K., & Petrosyan, L. A. (2010). Subgame consistent solutions for cooperative stochastic dynamic games. *Journal of Optimization Theory and Applications*, 145(3), 579–596.

Yeung, D. W. K., & Petrosyan, L. A. (2016). *Subgame Consistent Cooperation: A Comprehensive Treatise*, Singapore: Springer.

Yeung, D. W. K., & Petrosyan, L. A. (2025). A fair gain-sharing solution of cooperative dynamic games, *International Game Theory Review*, 27(03), 2550008. https://doi.org/10.1142/S0219198925500082.

Yeung, D. W. K., Petrosyan, L. A., & Zhang, Y. X. (2024). A fair-sharing value for cooperative dynamic network games. *Dynamic Games and Applications*, https://doi.org/10.1007/s13235-024-00562-x.

Yeung, D. W. K., Petrosyan, L. A., & Zhang, Y. X. (2025a). Managing Interconnected Production in a Dynamic Network: A Fair Gain-sharing Collaborative Solution, Contributions to Game Theory and Management, forthcoming 2025, Vol. XVII.

Yun, W. C., & Zhang, Z. X. (2006). Electric power grid interconnection in Northeast Asia. *Energy Policy*, 34(15), 2298–2309.

Zhang, X. P., Ou, M., Song, Y., & Li, X. (2017). Review of Middle East energy interconnection development. *Journal of Modern Power Systems and Clean Energy*, 5(6), 917–935.

Chapter 3

Application of Fair Gain-Sharing in Global Environmental Management

"Management, in the sense of employer, is merely the agent for the public, the stockholders and the employees. It is management's job to preserve the balance fairly between all these interests, that each may have his fair share without imperiling the continuity of the effort upon which the whole depends."

— **James F. Bell, III**

The rapid pace of industrial globalization and economic development has resulted in a swift rise in pollution levels. Collaboration represents the most effective approach to mitigate the detrimental impacts of individually rational non-cooperative behaviors in industrial production. Due to the disparities among the world's nations, certain countries may receive significantly less benefit than their contributions to other nations within a cooperative environmental management framework. A fair gain-sharing principle is essential for securing the consensus of all nations to engage in cooperation and for ensuring the sustainability of the scheme. This chapter presents a fair gain-sharing solution for global environmental management that meets the criteria of individual rationality, group optimality, time consistency, and the fair gain-sharing property that acknowledges the contributions of specific nations. Section 1 examines the gain-sharing problem in the context of global environmental management. Section 2 provides the theoretical underpinnings of a remedial solution. Section 3 develops an analytical framework and derives the non-cooperative equilibria and group-optimal outcomes. Section 4 establishes a dynamic fair-sharing cooperative solution for global environmental governance. An explication of the problems of the Conference of the Parties (COP) is given in Section 5. Section 6 offers a numerical example, while Section 7 provides the concluding remarks.

1. Gain-Sharing Problems in Global Environmental Management

After several decades of rapid technological advancement and economic growth, alarming levels of pollution and environmental degradation are emerging all over the world. Reports are portraying the situation as an industrial civilization on the verge of suicide, destroying its environmental conditions of existence, with people being held as prisoners on a runaway, catastrophe-bound train. Due to the geographical diffusion of pollutants, a unilateral response from a single nation or region is often inadequate. Non-cooperative actions by individual nations could lead to an "intertemporal tragedy of individually rational behaviors". Therefore, the formulation of a comprehensive cooperative scheme to address this globally devastating issue is urgently required. While collaboration in environmental management presents the most promising avenue for effective action, only very limited success has been achieved thus far. The existing multinational joint initiatives proposed in the COPs are unlikely to provide a sustainable solution, as they lack gain-sharing mechanisms that ensure all participants remain better off throughout the entire duration of the agreement.

Gain-sharing arrangements entail the allocation of benefits to collaborating participants in order to achieve a collective gain within a performance framework. In the realm of gain-sharing arrangements, decisions are influenced by arguments rooted in economic rationality and perceptions of fairness. These arrangements play a crucial role in determining whether participants choose to collaborate and ensuring the sustainability of the schemes. Numerous instances of collaborative scheme failures have been attributed to dissatisfaction with the gain-sharing arrangements. Gain-sharing arrangements are a vital component in establishing sustainable collaboration. Nevertheless, the disparities among participants regarding size, functionality, capability, resources, facilities, and growth potential pose significant challenges in creating gain-sharing arrangements that are perceived as fair by all involved parties.

We live in a global context where policymakers from various countries collaboratively formulate policy frameworks and initiatives aimed at establishing effective cooperation strategies to address worldwide challenges. Due to the asymmetry of nations, certain countries may believe that they receive significantly less benefit relative to their contributions in cooperative environmental controls. For example, a country that excels in the production of clean technology may face increased costs in developing such technology for the advantage of other nations. Conversely, a nation that generates substantial pollution would incur significant economic losses by reducing its environmentally harmful output, which would ultimately benefit all other countries. A fair principle of gain-sharing in cooperation is essential for securing the consensus of all nations to collaborate and for the long-term viability of the initiative. To achieve fair gain-sharing, the allocation of cooperative benefits must reflect the contributions made by each nation.

1.1. Asymmetries of the World Nations

The notion of fair gain-sharing is of great importance to our world, which consists of a large number of asymmetric nations. When considering GDP alone, the leading 10 nations globally are the United States with \$25.463 trillion, China with \$17.963 trillion, Japan with \$4.231 trillion, Germany with \$4.072 trillion, India with \$3.385 trillion, the United Kingdom with \$3.071 trillion, France with \$2.783 trillion, Russia with \$2.240 trillion, Canada with \$2.140 trillion, and Italy with \$2.010 trillion. The disparity among these nations reaches a maximum of over 12 times. In contrast, the 10 nations with the lowest GDP are Vanuatu with \$984 million, Saint Kitts & Nevis with \$962 million, St. Vincent & Grenadines with \$949 million, Samoa with \$832 million, Dominica with \$612 million, Sao Tome & Principe with \$547 million, Micronesia with \$427 million, Marshall Islands with \$280 million, Kiribati with \$223 million, and Tuvalu with \$60 million. The largest difference among these nations exceeds 16 times. Notably, the gap between the last nation in the first group

(Italy with $2.010 trillion) and the leading nation in the second group (Vanuatu with $984 million) is more than 2000 times, yet there are still over 150 nations whose GDP falls within this range of disparity.

The leading 10 countries in terms of GDP per capita include Monaco at $240,862, Liechtenstein at $187,267, Luxembourg at $128,259, Bermuda at $123,091, Ireland at $103,684, Switzerland at $99,994, the Cayman Islands at $96,073, Norway at $87,961, Qatar at $87,480, and Singapore at $84,734. Among these countries, the greatest disparity exceeds 2.8 times. The 10 countries with the lowest GDP per Capita are Mozambique at $645, Sudan at $606, North Korea at $590, Madagascar at $563, the Central African Republic at $529, Yemen at $465, Malawi at $464, Afghanistan at $411, South Sudan at $341, and Burundi at $321. The gap between the lowest nation in the first group (Singapore at $84,734 billion) and the highest nation in the second group (Mozambique at $645) is more than 130 times, yet there are still over 160 countries with GDP per capita within this range of disparity. Furthermore, the difference between the highest nation in the first group (Monaco at $240,862) and the lowest nation in the second group (Burundi at $321) exceeds 750 times.

On top of the measure of GDP, other factors are exhibiting the differences of the nations, including the stage of development. A developing country typically has a lower standard of living, an underdeveloped industrial sector, and a moderate to low human development index (HDI) score along with per capita income, yet it is undergoing economic growth. Generally, nations that do not fall into the categories of developed countries or failed states are considered developing nations. Nevertheless, it is important to note that certain developing countries may surpass some developed countries in terms of development. Nations that possess more advanced economies compared to other developing countries, yet have not fully exhibited the characteristics of a developed nation, are categorized as newly industrialized nations. In contrast, other developing countries that have shown consistent economic growth over time and demonstrate significant economic potential are referred to as emerging markets. The economies classified as big emerging markets (BEM) include Argentina, Brazil, China, Egypt, India, Indonesia, Mexico, Poland, Russia, South Africa, South Korea, and Turkey. There are also nations with an erratic development history, including many in Africa, Central America, and the Caribbean, a significant portion of the Arab World, and a few Southeast Asian countries like Laos and Cambodia. Additionally, some nations are plagued by long-term civil conflicts, significant breakdowns in the rule of law, or non-development-oriented dictatorships, such as Afghanistan, Haiti, Somalia, Myanmar, Iraq, and North Korea; these nations often face resource scarcity. Development requires a modern infrastructure, both physical and institutional, and a transition away from low-value-added sectors like agriculture and natural resource extraction. In contrast, developed nations typically have economic systems characterized by continuous, self-sustaining growth in the tertiary and quaternary sectors, along with high living standards.

1.2. Conference of the Parties: A Review

The most notable instrument for combating global environmental degradation is the Conference of the Parties (COP). It is commonly associated with the international meeting of the United Nations Framework Convention on Climate Change (UNFCCC) that concentrates on climate issues. The United Nations Climate Change Conferences are annual events conducted under the auspices of the United Nations Framework Convention on Climate Change (UNFCCC). These conferences act as the official gathering of the UNFCCC parties — known as the Conference of the Parties (COP) — to evaluate the advancements in addressing climate change. Since the mid-1990s, the COP has evolved into an annual conference aimed at reviewing progress in tackling climate change.

1995: COP 1, Berlin, Germany

The central issue was the adequacy of individual country commitments, resulting in a mandate to begin a process toward individual country action for the period beyond 2000. Conference delegates did not reach a consensus on the Rules of Procedure.

1996: COP 2, Geneva, Switzerland

The ministerial declaration was noted (but not adopted). It accepted the scientific findings on climate change proffered by the Intergovernmental Panel on Climate Change (IPCC) and called for "legally binding mid-term targets".

1997: COP 3, Kyoto, Japan

It outlined the greenhouse gas emissions reduction obligation for Annex I countries, along with what came to be known as Kyoto mechanisms, such as emissions trading, clean development mechanism, and joint implementation. The United States would be required to reduce its total emissions by an average of 7% below 1990 levels. The US Congress did not ratify the treaty after Clinton signed it. The Bush administration explicitly rejected the protocol in 2001.

1998: COP 4, Buenos Aires, Argentina

It had been expected that the remaining issues unresolved in Kyoto would be finalized at this meeting. However, the complexity and difficulty of finding agreement on these issues proved insurmountable. The parties adopted a two-year "Buenos Aires Plan of Action" (BAPA) to advance efforts for implementing the Kyoto Protocol to be completed by 2000.

1999: COP 5, Bonn, Germany

It was primarily a technical meeting and did not reach any major conclusions. The 165 conference delegates continued their work toward fulfilling the Buenos Aires Plan of Action (BAPA) adopted at COP 4.

2000: COP 6, The Hague, Netherlands

COP 6 was intended to complete work on the two-year Buenos Aires Plan of Action (BAPA) agreed upon at COP 4. Despite some compromises agreed between the United States and some EU countries, the EU countries as a whole, led by Denmark and Germany, rejected the compromise positions, and the talks in The Hague collapsed.

2001: COP 6-2, Bonn, Germany

This meeting took place after George W. Bush had become the President of the United States and had rejected the Kyoto Protocol in March 2001. The United States delegation to this meeting declined to participate in the negotiations and chose to take the role of observer at the meeting. The agreements reached by the other parties covered areas concerning (i) flexible mechanisms, (ii) carbon sinks, (iii) compliance, and (iv) financing. A number of operational details attendant upon these decisions remained to be negotiated and agreed upon.

2001: COP 7, Marrakech, Morocco

Negotiators wrapped up the work on the Buenos Aires Plan of Action, finalizing most of the operational details and setting the stage for nations to ratify the Kyoto Protocol. The United States delegation maintained its observer role, declining to participate actively in the negotiations. The main decisions included operational rules for international emissions trading, a compliance regime, and accounting procedures for the flexibility mechanisms.

2002: COP 8, New Delhi, India

COP 8 called for efforts by developed countries to transfer technology and minimize the impact of climate change on developing countries. Russia showed hesitation, stating that it needed more time for consideration.

2003: COP 9, Milan, Italy

The parties agreed to use the Adaptation Fund established primarily to support developing countries to better adapt to climate change. The fund would also be used for capacity-building through technology transfer. The parties also agreed to review the first national reports submitted by 110 countries.

2004: COP 10, Buenos Aires, Argentina

The conference discussed the progress made since the first COP 10 years ago and its future challenges, with special emphasis on climate change mitigation and adaptation. The Buenos Aires Plan of Action was adopted to help developing countries better adapt to climate change. The parties also discussed the post-Kyoto mechanism on how to allocate emission reduction obligations after 2012.

2005: COP 11/CMP 1, Montreal, Canada

The event, which hosted more than 10,000 delegates, marked the entry into force of the Kyoto Protocol. The Montreal Action Plan was an agreement to extend the life of the Kyoto Protocol beyond its 2012 expiration and negotiate deeper cuts in greenhouse-gas emissions.

2006: COP 12/CMP 2, Nairobi, Kenya

Due to the concerns over economic costs and possible losses of competitiveness, the majority of the discussions avoided the issue of reducing emissions. The conference adopted a five-year plan of work to support climate change adaptation by developing countries and agreed on the procedures of the Adaptation Fund. It also agreed to improve the projects for the clean development mechanism.

2007: COP 13/CMP 3, Bali, Indonesia

Agreement on a timeline and structured negotiation on the post-Kyoto framework in 2012 was achieved with the adoption of the Bali Action Plan. The Working Group on Long-term Cooperative Action under the convention was established as a new subsidiary body to conduct the negotiations aimed at enhancing the implementation of the convention up to and beyond 2012. The negotiations took place during COP 14 in 2008.

2008: COP 14/CMP 4, Poznań, Poland

The delegates agreed on principles for the financing of a fund to help the poorest nations cope with the effects of climate change. They also approved a mechanism to incorporate forest protection into the efforts of the international community to combat climate change. Negotiations on a successor to the Kyoto Protocol were the primary focus of the conference.

2009: COP 15/CMP 5, Copenhagen, Denmark

The overall goal of the conference was to establish an ambitious global climate agreement for the period after 2012. Many industrialized countries were reluctant to fulfill commitments under the Kyoto Protocol. Some countries, including the US, decided to put off the difficult task of reaching a climate change agreement. Instead, they arrived at a less specific binding agreement that left the more difficult issues for the future. The conference did not achieve a binding agreement for long-term action. COP 15 referred to a collective commitment by developed countries for new and additional resources, including forestry and investments through international institutions.

2010: COP 16/CMP 6, Cancún, Mexico

The outcome of the summit was an agreement adopted by the states' parties that called for the US$100 billion per annum "Green Climate Fund" and a "Climate Technology Centre" and network. However, the funding of the "Green Climate Fund" was not agreed upon.

All parties recognized that climate change was an urgent and potentially irreversible threat to human societies and the planet and thus required to be urgently addressed by all parties. It recognized the IPCC Assessment Report goal of a maximum 2°C global warming, and all parties should take urgent action to meet this goal.

2011: COP 17/CMP 7, Durban, South Africa

The conference agreed to start negotiations on a legally binding deal comprising all countries, to be adopted in 2015, governing the period post 2020. There was also progress regarding the creation of a Green Climate Fund (GCF), for which a management framework was adopted. The fund was to distribute US$100 billion per year to help poor countries adapt to climate impacts. Scientists warned that the deal was not sufficient to avoid global warming beyond 2°C and that more urgent actions were needed.

2012: COP 18/CMP 8, Doha, Qatar

The conference produced a package of documents collectively titled The Doha Climate Gateway, which made amendments to the Kyoto Protocol. The conference made little progress toward the funding of the Green Climate Fund. Russia, Belarus, and Ukraine objected at the end of the session.

2013: COP 19/CMP 9, Warsaw, Poland

The most prominent result was the adoption of the Warsaw Framework for REDD-plus. The conference also established the Warsaw International Mechanism (WIM) for Loss and Damage associated with Climate Change Impacts (Loss and Damage Mechanism), to address loss and damage associated with impacts of climate change.

2014: COP 20/CMP 10, Lima, Peru

The conference hosted the 20th yearly session of the COP. Negotiations in Lima focused on matters prepared to advance toward an agreement in Paris at COP 21 in 2015, including elaboration of the information and process required for the submission of intended nationally determined contributions. It adopted the "Lima Call for Climate Action", which set in motion the negotiations in the coming year toward a 2015 agreement.

2015: COP 21/CMP 11, Paris, France

Negotiations resulted in the adoption of the Paris Agreement, governing climate change reduction measures from 2020. The adoption of this agreement ended the work of the Durban platform established during COP 17. The agreement entered into force (and thus became fully effective) when the threshold for adoption was reached, with over 55 countries representing at least 55% of the world's greenhouse gas emissions ratifying the agreement.

2016: COP 22/CMP 12/CMA 1, Marrakech, Morocco

A focal issue of COP 22 was on water scarcity, water cleanliness, and water-related sustainability, a major problem in the developing world, including many African states. Another focal issue was the need to reduce greenhouse emissions and utilize low-carbon energy sources. There was a call for the transformation of the global economy in all sectors to achieve a low-emissions global economy.

2017: COP 23/CMP 13/CMA 1-2, Bonn, Germany

The purpose of the conference was to discuss and implement plans about combating climate change, including the details of how the Paris Agreement will work after it enters into force in 2020. COP23 concluded with the "Fiji Momentum for Implementation", which outlined the steps that need to be taken in 2018 to make the Paris Agreement operational and launched the Talanoa Dialogue — a process designed to help countries enhance and implement their Nationally Determined Contributions by 2020.

2018: COP 24/CMP 14/CMA 1-3, Katowice, Poland

The conference agreed on rules to implement the Paris Agreement — the rulebook on how governments will measure and report on their emissions-cutting efforts. Due to the difficulty to reach agreement between parties, some difficult questions, such as ways to scale up existing commitments on cutting emissions, ways to provide financial help for poor countries, wording that does not allow double counting, and whether countries are doing enough to cut their emissions, were postponed to the next conference.

2019: COP 25/CMP 15/CMA 2, Madrid, Spain

The results of the conference were disappointing at a time when climate action and concrete measures were considered urgent.

The decisions about the carbon market and emissions cuts were delayed to the next climate conference in Glasgow. The United States, Russia, India, China, Brazil, and Saudi Arabia were the main opponents of these measures.

2021: COP 26/CMP 16/CMA 3, Glasgow, United Kingdom

This conference led to the development of the Accelerating to Zero coalition to accelerate the phase-out of fossil fuel vehicles and the Glasgow Climate Pact to phase down the use of coal-fired power stations. The US and China agreed on a framework to reduce carbon emissions by cooperating on measures to lower the use of methane, phase out the use of coal, and increase protection of forests. A group of 22 countries, including China and India, asked for the commitment to mitigation to be entirely removed from the draft text, as they argued that developing countries should not be held to the same deadlines as wealthier nations.

2022: COP 27/CMP 17/CMA 4, Sharm El Sheikh, Egypt

The conference led to an agreement on loss and damage, under which rich countries could compensate poor countries for damage caused by climate change, and a loss and damage fund was agreed to provide funding to countries that are most vulnerable and affected by climate change. No commitment to phase out fossil fuels was made. Although they have focused on "low-emission energy", the failure to move away from fossil fuels would continue to be a source of greenhouse gas emissions.

2023: COP 28/CMP 18/CMA 5, Dubai, UAE

A final compromised agreement between the countries involved had been reached. The deal commits all signatory countries to move away from carbon energy sources "in a just, orderly and equitable manner" to mitigate the worst effects of climate change and reach net zero by the year 2050. The UAE Consensus was the first in the history of COP summits to explicitly mention the need to shift away from every type of fossil fuel, but it still received criticism due to the lack of a clear commitment to either fossil fuel phase-out or phase-down. China and India did not sign the pledge to triple their output of renewable energy and committed to coal power instead.

2024: COP 29/CMP 19/CMA 6, Baku, Azerbaijan

At the conference, climate finance was considered a central topic, focusing on scaling up resources for developing countries to address climate impacts and transition to low-carbon economies. Delegations reached an agreement for a deal that would oblige developed nations to make a $300 billion annual climate finance pledge by 2035 in order to help developing countries get access to wider economic resources to tackle emissions and climate disasters. However, poorer nations criticized the amount as insufficient, with many pushing for a minimum of $500 billion target. While the deal was seen as a small step, some critics called it a betrayal, especially due to the lack of mandatory contributions for emerging economies like China.

2025: COP30, Belém, Brazil

At the conference, a review of the Paris Agreement showed that current efforts are not sufficient to keep warming below 1.5° C. Developed nations pledged increased contributions beyond the $100 billion annual goal directed to adaptation and loss, and damage funds, addressing vulnerable nations' demands. Disagreements on methodology and accountability stalled progress. The United States did not send an official delegation, raising concerns about global leadership. Developing nations insist on equity and historical responsibility, while developed nations resist binding obligations under the Paris Agreement. Sharp divides emerged between blocs, such as the Like-Minded Developing Countries (LMDCs), Alliance of Small Island States (AOSIS), and the EU, over how finance should be integrated into the final text. Disagreements on methodology and accountability stalled progress. COP30 ended without a consensus on the most urgent issues. COP30 set the stage for COP31 in 2026, where nations must present enhanced implementation plans.

In conclusion, COP is expected to be a long-term framework for the collaboration of nations aimed at addressing environmental degradation. Recent conferences have highlighted that the primary obstacle to achieving a successful agreement lies in the differing views on the allocation of contributions from affluent nations and the distribution of subsidies to developing countries. Around the time of COP 29, UN Secretary-General António Guterres issued a compelling appeal to world leaders and negotiators, urging them to resolve their differences and secure a landmark climate finance agreement. He stated, "Failure on climate finance is 'not an option'... and expressed hope for a more ambitious outcome regarding both finance and mitigation to tackle the significant challenges we face." COP30 showed that the fight against environmental degradation is far from being won, and global climate diplomacy is still fragile and not moving forward significantly.

2. Theoretical Underpinnings of a Reference Solution

The COPs primarily concentrated on reducing emissions, permit trading, and regulatory measures. Though collaboration in environmental governance presents the greatest potential for effective actions, limited success has been noted due to several critical issues that remain unresolved within the COP cooperation framework. These critical issues include the following:

(i) The efficacy of the policy instruments in addressing the current environmental challenges. The reduction of pollutant emissions is insufficient to enhance the environment in the context of rapid industrialization, global economic growth, and environmental degradation. A strategy to address the escalating demand resulting from economic growth over time is lacking.

(ii) There is no assurance that the participating countries will not experience a decline in their well-being (in comparison to their non-cooperative state) during the cooperation period.

(iii) The economic feasibility of the policy for all participants after entering the cooperation framework is not adequately addressed. Reductions in pollutant emissions may lead to significant economic challenges for low-income countries.

(iv) There is no guarantee that the initial positions that nations are entitled to hold at the outset will be preserved throughout all subsequent phases of the cooperation framework. It is possible that some participants may see their share of the cooperative gains diminish significantly in later stages, prompting them to withdraw from the scheme (which could destabilize the dynamic cooperation framework).

(v) The current cooperation framework does not achieve the maximized joint payoff and fails to capture all potential gains, resulting in no incentive for participants to deviate from the agreed-upon solution plan (or to negotiate for an alternative plan) in order to access the unexploited gains.

(vi) Cooperative development of clean technology and phasing out environmentally degrading techniques have not been seriously implemented. Adoption of environment-preserving and sustainable technology plays a central role in solving the problem effectively. Nevertheless, the majority of current environmentally friendly technologies are more expensive than traditional technologies. There is a lack of sufficient incentives for well-developed countries to embrace clean technology, and it is economically unfeasible for less developed nations. Many existing environmentally friendly methods, such as solar energy, recycling processes, sustainable renewable resources, wind energy, ocean energy, and motor system optimization, are not yet fully optimized for cost efficiency. A joint effort in the development of clean technology and coordinated innovation within a collaborative framework could produce beneficial externalities, particularly in terms of lowering developmental costs.

On the other hand, the fair gain-sharing solution in Chapter 2 offers the potential to employ a broader range of strategies that are pertinent to the effective cooperative management of environmental challenges. These strategies include the following:

(i) A comprehensive set of policy instruments, which encompasses levies on industrial pollution, royalties on natural resources, coordinated efforts for pollutant abatement, collaborative development of environmentally sustainable technology, and transfer payments aimed at achieving an optimal cooperation solution, can be formulated.

(ii) The effective joint development of environmentally sustainable technology alongside an optimal pattern for adopting both conventional and clean technologies over time is important. The collaborative development of such technology harnesses the positive externalities associated with research and development (R&D) and innovation, which cannot be realized through individual nation efforts.

(iii) A cooperative maximization program can be designed to yield the optimal (highest) joint payoff, ensuring that all gains are captured so that no participants have the incentive to deviate from the agreed-upon solution plan in order to exploit unexploited gains.

(iv) A guarantee should be given that the cooperative payoffs for the participating nations will consistently exceed their non-cooperative payoffs at every stage of the cooperation along the optimal cooperation trajectory. The fundamental aspect of cooperation — individual rationality — is assured throughout the duration of the cooperation.

(v) A financially sustainable plan can be offered to each participating country that provides a cooperative payoff exceeding its non-cooperative payoff via transfer payments.

(vi) A subgame-consistent method for distributing payoffs should guarantee the preservation of the initially agreed optimality principle in gain-sharing throughout the cooperation period. This eliminates the motivation for participating countries to withdraw from the cooperation scheme if circumstances change in the future.

(vii) A fair gain-sharing arrangement must represent the contributions of the nations to the cooperative payoff.

3. Game Formulation and Equilibria of a Reference Solution

To present a reference solution for solving the problems of COP, we first provide the analytical framework and characterize the non-cooperative equilibrium and group optimal outcome.

3.1. Nations' Strategies, State Variables, and Evolution Dynamics

The entire set of strategies of all participating nations is defined as $\underline{u}_k = (u_k^1, u_k^2, \ldots, u_k^n)$. This set of strategies includes the following: resource extraction, clean transportation, industrial production, investment in green technology, technology transfer and innovation, R&D, subsidies and investments, government spending, taxes, pollution abatement activities, ecosystem improvement activities, and various environmental regulatory policies.

Major state variables include air and water pollution, the condition of the ecosystems, the level of green technology, the level of productive capital, the level of biodiversity, government debt, private capital, and public capital. In particular, we can express the vector of states as $x_k = (x_k^1, x_k^2, \ldots, x_k^m)$.

The evolution dynamics of the state can be expressed as

$$\begin{pmatrix} x_{k+1}^1 = f_k^1(x_k, \underline{u}_k) \\ x_{k+1}^2 = f_k^2(x_k, \underline{u}_k) \\ \vdots \\ x_{k+1}^m = f_k^m(x_k, \underline{u}_k) \end{pmatrix} \equiv x_{k+1} = f_k(x_k, u_k). \tag{1}$$

3.2. Objectives and Payoffs

The objectives of the cooperation scheme include the following: (i) air quality improvement, (ii) improving waste management, (iii) minimizing toxic exposure, (iv) good water quality, (iv) reducing noise pollution, (v) resource conservation, (vi) efficiency in energy use, (vii) sustainable ecosystem, (viii) biodiversity, (ix) maintaining employment and income, (x) government revenue, and (xi) economic growth. Concerns will be given to all participating nations with respect to the monetary weights of their objectives.

The gain of nation i in period k depends on (x_k, u_k) and is expressed as $\sum_{\varsigma=1}^{\Theta_i} w_k^{i(\varsigma)} g_k^{i(\varsigma)}(x_k, \underline{u}_k)$ if nation i has Θ_i objectives, where $g_k^{i(\varsigma)}(x_k, \underline{u}_k)$ is the ςth objective and $w_k^{i(\varsigma)}$ is the corresponding monetary weight. If $\Theta_i = 1$, the gain of nation i in period k reduces to $w_k^{i(1)} g_k^{i(1)}(x_k, \underline{u}_k)$. The terminal payoff of nation i at period $T+1$ depends on x_{T+1} and is expressed as $q_{T+1}^i(x_{T+1})$.

3.3. Non-cooperative Equilibria

The payoff of nation $i \in \{1, 2, \ldots, n\} \equiv N$ over the period 1 to period T can be expressed as

$$\sum_{k=1}^{T} \sum_{\varsigma=1}^{\Theta_i} w_k^{i(\varsigma)} g_k^{i(\varsigma)}(x_k, \underline{u}_k) \delta_1^k + q_{T+1}^i(x_{T+1}) \delta_1^{T+1}. \tag{2}$$

Under non-cooperation, it becomes a general n-nation game:

$$\max_{\substack{u_k^i \\ k\in\{1,2,\ldots,T\}}} \left\{ \sum_{k=1}^{T} \sum_{\varsigma=1}^{\Theta_i} w_k^{i(\varsigma)} g_k^{i(\varsigma)}(x_k, \underline{u}_k)\delta_1^k + q_{T+1}^i(x_{T+1})\delta_1^{T+1} \right\}, \quad i \in N, \tag{3}$$

subject to

$$x_{k+1} = f_k(x_k, \underline{u}_k), \quad x_1 = x_1^0. \tag{4}$$

Invoking Theorem 2.1 in Chapter 2, a feedback Nash equilibrium of the game (3)–(4) can be characterized by the following corollary.

Corollary 3.1. *A set of strategies $\{u_k^{i**}$ for $k \in \{1, 2, \ldots, T\}$ and $i \in N\}$ provides a feedback Nash equilibrium solution to the game (3)–(4) if there exist functions $V^i(kx)$ for $k \in K$ and $i \in N$, such that the following recursive relations are satisfied:*

$$V^i(T+1, x) = q_{T+1}^i(x)\delta_1^T, \tag{5}$$

$$\begin{aligned} V^i(k, x) = \max_{u_k^i} \Bigg\{ & \sum_{\varsigma=1}^{\Theta_i} w_k^{i(\varsigma)} g_k^{i(\varsigma)}[x, u_k^{1**}, u_k^{2**}, \ldots, u_k^{i-1**}, \\ & u_k^i, u_k^{i+1**}, \ldots, u_k^{n**}]\delta_1^k + V^i[k+1, f_k(x, u_k^{1**}, u_k^{2**}, \ldots, \\ & u_k^{i-1**}, u_k^i, u_k^{i+1**}, \ldots, u_k^{n**})\Bigg\} \end{aligned} \tag{6}$$

for $i \in N$ and $k \in \{1, 2, \ldots, T\}$.

Cooperation will stop if any nation refuses to continue with the scheme, and they will return to the non-cooperative situation.

3.4. Group Optimal Solution

Now, consider the case when the nations agree to cooperate and maximize their payoff by solving the problem

$$\max_{\substack{\underline{u}_k^j, j\in N \\ k\in\{1,2,\ldots,T\}}} \left\{ \sum_{j=1}^{n} \left[\sum_{k=1}^{T} \sum_{\varsigma=1}^{\Theta_j} w_k^{j(\varsigma)} g_k^{j(\varsigma)}(x_k, \underline{u}_k)\delta_1^k + q_{T+1}^j(x_{T+1})\delta_1^{T+1} \right] \right. \tag{7}$$

subject to dynamics (1).

Invoking Theorem 2.2 in Chapter 2, an optimal solution to the control problems (1) and (7) can be characterized by the following corollary.

Corollary 3.2. *A set of strategies* $\{u_k^{i*}$ *for* $k \in \{1, 2, \ldots, T\}$ *and* $i \in N\}$ *provides an optimal solution to the problems (1) and (7) if there exist functions* $W^N(k, x)$ *for* $k \in \{1, 2, \ldots, T\}$*, such that the following recursive relations are satisfied:*

$$W^N(T+1, x) = \sum_{j=1}^{n} q_{T+1}^j(x)\delta_1^T, \tag{8}$$

$$W^N(k, x) = \max_{u_k^1, u_k^2, \ldots, u_k^n} \left\{ \sum_{j=1}^{n} \sum_{\varsigma=1}^{\Theta_j} w_k^{j(\varsigma)} g_k^{j(\varsigma)}[x_k, u_k^1, u_k^2, \ldots, u_k^n]\delta_1^k \right.$$
$$\left. + W^N[k+1, f_k(x_k, u_k^1, u_k^2, \ldots, u_k^n)] \right\} \tag{9}$$

for $k \in \{1, 2, \ldots, T\}$.

Substituting the optimal control $\{u_k^{i*}$ for $k \in \{1, 2, \ldots, T\}$ and $i \in N\}$ into the state dynamics (1), one can obtain the dynamics of the cooperative trajectory as

$$x_{k+1} = f_k(x_k, u_k^{1*}, u_k^{2*}, \ldots, u_k^{n*}). \tag{10}$$

We use $\{x_k^*\}_{k=1}^T$ denote the solution to (9), which is the optimal cooperative path.

4. A Dynamic Fair-Sharing Cooperative Solution

In this section, we first derive a dynamically stable fair gain-sharing solution, followed by a mechanism for payoff distribution that enables the realization of the derived gain-sharing solution. Considering the asymmetry in payoffs and the sizes of nations, it is crucial to develop a fair distribution scheme that reflects the contributions made by the nations. We first calculate the marginal contribution of a nation to the total joint payoff in Section 4.1. Then, we provide a fair gain-sharing solution that reflects the contributions of the nations in Section 4.2.

4.1. Marginal Contributions of Nations

To derive a dynamic fair gain-sharing cooperative solution, we first calculate the additional value to the joint payoff of cooperating nations if an additional nation, say nation i, is included in the cooperation pack. For notational convenience, we use Ni to denote the set of nations without nation i. The addition to the maximized joint payoff by including nation i is the difference between the maximized joint payoff and the maximized joint payoff without nation i. Since the addition to the maximized payoff by nation i is the difference between the maximized joint payoff of the collaborating scheme and the maximized payoff without i, we have to solve the game:

$$\max_{\substack{u_k^j, j \in N\backslash i \\ k \in \{1,2,\ldots,T\}}} \left\{ \sum_{j \in N\backslash i} \left[\sum_{k=1}^{T} \sum_{\zeta=1}^{\Theta_j} w_k^{j(\zeta)} g_k^{j(\zeta)} (x_k, \underline{u}_k) \, \delta_1^k + q_{T+1}^j (x_{T+1}) \, \delta_1^{T+1} \right] \right\} \tag{11}$$

$$\max_{\substack{u_k^i, \bar{u}_k^i \\ k \in \{1,2,\ldots,T\}}} \left\{ \sum_{k=1}^{T} \sum_{\zeta=1}^{\Theta_i} w_k^{i(\zeta)} g_k^{i(\zeta)} (x_k, \underline{u}_k) \, \delta_1^k + q_{T+1}^i (x_{T+1}) \, \delta_1^{T+1} \right\} \text{ for } i \in N \tag{12}$$

subject to dynamic (1).

Invoking Theorem 2.1 in Chapter 2, the non-cooperative equilibrium of the game (1) and (11)-(12) can be characterized by the following corollary.

Corollary 4.1. *A set of strategies* $\{\hat{u}_k^i$ *for* $k \in \{1, 2, \ldots, T\}$ *and* $\{u_k^{(N\backslash i)j}$ *for* $k \in \{1, 2, \ldots, T\}$ *and* $j \in N\backslash i\}$ *constitutes a set of feedback Nash equilibria of the game (1) and (11)-(12) if there exists functions* $W^{N\backslash i}(k, x)$ *and* $\hat{V}^i(k, x)$, *which satisfy the following recursive equations:*

$$W^{N\backslash i}(T+1, x) = \sum_{j \in N\backslash i} q_{T+1}^j (x_{T+1}) \, \delta_1^{T+1} \tag{13}$$

$$W^{N\backslash i}(k, x) = \max_{u_k^j, j \in N\backslash i} \left\{ \sum_{j \in N\backslash i} \sum_{\varsigma=1}^{\Theta_j} w_k^{j(\varsigma)} g_k^{j(\varsigma)} (x, u_k^1, u_k^2, \ldots, \right.$$

$$u_k^{i-1}, \hat{u}_k^i, u_k^{i+1}, \ldots, u_k^n) \delta_1^k$$

$$+ W^{N\backslash i}[k+1, f_k(x, u_k^1, u_k^2, \ldots, u_k^{i-1}, \hat{u}_k^i, u_k^{i+1}, \ldots, u_k^n)]\Bigg\} \tag{14}$$

for $k \in \{1, 2, \ldots, T\}$;

$$\widehat{V}^i(T+1, x) = q_{T+1}^i(x)\delta_1^{T+1}, \tag{15}$$

$$\hat{V}^i(k,x) = \max_{u_k^i}\left\{\sum_{\zeta=1}^{\Theta_i} w_k^{i(\zeta)} g_k^{i(\varsigma)}(x, u_k^{(N\backslash i)1}, u_k^{(N\backslash i)2}, \ldots, u_k^{(N\backslash i)i-1}, u_k^i, u_k^{(N\backslash i)i+1}, \ldots, u_k^{(N\backslash i)n})\delta_1^k + \hat{V}^i[k+1, f_k(u_k^{(N\backslash i)1}, u_k^{(N\backslash i)2}, \ldots, u_k^{(N\backslash i)i-1}, u_k^i, u_k^{(N\backslash i)i+1}, \ldots, u_k^{(N\backslash i)n})]\right\} \tag{16}$$

for $k \in \{1, 2, \ldots, T\}$.

The characterization of the game equilibrium of the game in Corollary 4.1 allows one to derive the joint payoff of nations if nation i is excluded, i.e., $W^{N\backslash i}(k,x)$.

4.2. Fair Gain-Sharing Solution

Invoking equation (11), a fair gain-sharing weight for measuring nation i's share of contribution to the maximized joint payoff can be obtained as

$$\frac{W^N(k,x) - W^{N\backslash i}(k,x)}{\sum_{\zeta=1}^{n}\left[W^N(k,x) - W^{N\backslash\varsigma}(k,x)\right]} \text{ for nation } i \in N. \tag{17}$$

The weight (17) reflects the relative size of nation i's contribution to the cooperative gain. A dynamically stable fair gain-sharing value can be obtained as

$$\xi^i(k, x_k^*) = \frac{W^N(k, x_k^*) - W^{N\backslash i}(k, x_k^*)}{\sum_{\varsigma=1}^{n}[W^N\left(k, x_k^*\right) - W^{N\backslash\varsigma}(k, x_k^*)]}$$

$$\times \left[W^N(k, x_k^*) - \sum_{j=1}^{n} V^j(k, x_k^*) \right] + V^i(k, x_k^*) \qquad (18)$$

for $i \in N$ along the optimal cooperative trajectory $\{x_k^*\}_{k=1}^T$.

Following Theorem 4.3 in Chapter 2, a payment

$$\beta_k^i(x_k^*) = (\delta_1^k)^{-1} \left[\xi^i(k, x_k^*) - \xi^i(k+1, f_k(x_k^*, \underline{u}_k^*)) \right] \qquad (19)$$

given to nation i in stage k will lead to the realization of the fair gain-sharing dynamic cooperative value in (18).

Since under cooperation, all nations would use the cooperative strategies and the payoff that nation i will directly receive at stage k along the cooperative trajectory equals $\sum_{\varsigma=1}^{\Theta_i} w_k^{i(\varsigma)} g_k^{i(\varsigma)}(x_k^*, \underline{u}_k^*)$. Therefore, a side payment

$$\pi_k^i(x_k^*) = \beta_k^i(x_k^*) - \sum_{\zeta=1}^{\Theta_i} w_k^{i(\zeta)} g_k^{i(\zeta)}(x_k^*, \underline{u}_k^*) \qquad (20)$$

has to be given to nation $i \in N$ in period for $k \in \{1, 2, \ldots, T\}$ to yield the cooperative payoff $\xi^i(k, x_k^*)$.

5. Explication in COP Resolution Plan

Fair gain-sharing cooperation represents a reliable commitment to effectively address the issues of disagreement within the COP regarding the contributions of affluent nations and the allocation of subsidies to less wealthy nations. The top four most crucial issues that UN Secretary-General António Guterres had highlighted in COP29 were as follows: (i) clock is ticking, (ii) a clear path forward, (iii) an investment, not a handout, and (iv) a call for unity. These issues are still not being resolved in COP30. The fair gain-sharing solution responds to these as follows:

(i) Clock is ticking:

UN's Call: "While progress has been made and areas of convergence are emerging, significant differences still remain. But without decisive action, the consequences could ripple far beyond. Failure might jeopardize both near-term action and ambition in the preparation of new national climate action plans, and it could accelerate the approach of irreversible climate tipping points."

The fair gain-sharing cooperative solution can provide an immediate agenda for the UN to be discussed in the upcoming COP.

(ii) A clear path forward:

UN's Call: "The critical need for an ambitious new climate finance goal: a comprehensive financial package designed to mobilize resources for developing countries, enabling them to implement climate action plan target. The importance of financing initiatives that support nations in transitioning to clean, affordable energy while reducing emissions. The significance of the COP agreement is more than a mere negotiation. This is a COP to deliver justice in the face of climate catastrophe."

The fair gain-sharing cooperative solution can provide a comprehensive financial package to mobilize resources for both the developed and developing countries. The cooperative plan is guided by optimal actions of all nations and fair gain-sharing (and not just negotiation). It entails justice in the form that the gain-sharing fairly reflects the contributions of individual nations. It provides a possible clear path forward to avoid a climate catastrophe with group optimal, individual rationality, and time consistency. The solution also expanded the rather restrictive set of cooperative strategies to a large set of strategies, which include industrial production, investment in technologies, technology transfer and innovation, R&D, subsidies and investments, pollution abatement activities, and ecosystem improvement activities.

(iii) An investment, not a handout:

UN's Call: "Challenging the notion of climate finance as a form of charity and it is a critical investment in the planet's future. It is a down payment on a safer, more prosperous future for every nation on Earth."

The distribution of payoffs in the gain-sharing cooperative solution is determined by contributions and not by handouts. The efforts made by collaborating nations in reducing polluting production, implementing clean technologies, and safeguarding the ecosystem justify the need for climate financing. The dynamic optimization process ensures a viable, safer, and more prosperous future for every nation on Earth.

(iv) A call for unity:

UN's Call: "Recognizing the geopolitical divisions that could undermine progress, leaders and negotiators are urged to 'soften hard lines', navigate their differences and "keep eyes on the bigger picture. Appealing for unity and reminding all parties of what is at stake - a liveable planet for future generations... This is not a zero-sum game."

The fair gain-sharing cooperative solution is a global dynamic optimal scheme that represents the best possibility for maintaining a livable planet for future generations. The distribution of the gains is proportional to the nations' contributions. The agreed-upon cooperative actions lead to higher cooperative gains (over non-cooperative gains) for all nations. Hence, they do not need to hold on to hard lines, and there exist incentives to unite to pursue this endeavor that benefits every one of them. Finally, the solution is not a zero-sum game.

6. A Numerical Example of International Climate Control

Consider a T-periods and n-nations international political encounter in environmental management. The pollution stock in period k is x_k, and the dynamics of pollution accumulation is governed by the difference equation

$$x_{k+1} = x_k + \sum_{i=1}^{n} a_k^i \bar{q}_k^i - \sum_{i=1}^{n} b_k^i u_k^i \left(x_k\right)^{1/2} - \gamma_k x_k, x_1 = x_1^0, \tag{21}$$

where $a_k^i \bar{q}_k^i$ is the amount of pollution created by a unit of nation i 's polluting output, $\bar{q}_k^i, b_k^i u_k^i \left(x_k\right)^{1/2}$ is the amount of pollution removed by u_k^i units of abatement effort from nation i, and γ_k is the natural rate of decay of the pollutant stock.

The level of green technology in period k is ϑ_k, and the evolution of the level of clean technology is governed by the difference equation

$$\vartheta_{k+1} = \vartheta_k + \sum_{i=1}^{n} z_k^i w_k^i - \lambda_k \vartheta_k, \vartheta_1 = \vartheta_1^0, \tag{22}$$

where $\boldsymbol{z_k^i w_k^i}$ is the advancement in clean technology brought about by $\overline{\boldsymbol{w}}_{\boldsymbol{k}}^{\boldsymbol{i}}$ in period $\boldsymbol{k}$ and $\boldsymbol{\lambda_k}$ is the rate of obsolescence of the level of green technology.

Each nation seeks to maximize the sum of the discounted payoffs over the T-periods plus the terminal appraisal. In particular, nation $i \in N$ would seek to maximize the objective

$$\sum_{k=1}^{T}[\alpha_k^i \bar{q}_k^i - \beta_k^i (\bar{q}_k^i)^2 - c_k^i \bar{q}_k^i + 2\varpi_k^i (\vartheta_k q_k^i)^{1/2} - \varepsilon_k^i q_k^i + \chi_k w_k^i - (w_k^i)^2 - c_k^{(a)i}(u_k^i)^2 - h_k^i x_k]\delta^{k-1} + (h_{T+1}^i x_{T+1} + y_{T+1}^i \vartheta_{T+1} + Q_{T+1}^i)\delta^T \tag{23}$$

where $\alpha_k^i \bar{q}_k^i - \beta_k^i (\bar{q}_k^i)^2 - c_k^i \bar{q}_k^i$ is the net economic profit in the polluting industrial sector that generates greenhouse gas, $2\varpi_k^i (\vartheta_k q_k^i)^{1/2} - \varepsilon_k^i q_k^i$ is the net economic profit in the green industrial sector, $\chi_k^i w_k^i - (w_k^i)^2$ is the net economic profit in the green technology development sector, $c_k^{(a)i}(u_k^i)^2$ is the cost of the government's abatement effort u_k^i, $h_k^i x_k$ is the damage cost of the amount of pollution, $(h_{T+1}^i x_{T+1} + y_{T+1}^i \vartheta_{T+1} + Q_{T+1}^i)$ is the terminal payoff of nation i in period $T+1$, and δ^{k-1} is the discount factor from stage 1 to stage k.

The problem of maximizing the nation's objective (23) with controls subject to pollution dynamics (21) and clean technology development dynamics (22) is an n-nation dynamic game.

6.1. Non-cooperative Equilibrium

The game equilibrium payoffs and strategies of the nations are expressed in the following proposition.

Proposition 6.1. (1) *The game equilibrium payoffs of the nations are*

$$V^i(k, x, \vartheta) = (A_k^i x + B_k^i \vartheta + C_k^i)\delta^{k-1},$$

where

$$A_k^i = -c_k^{(a)i}\left(\frac{(b_k^i)^2}{2c_k^{(a)i}} A_{k+1}^i \delta\right)^2 - h_k^i + A_{k+1}^i \left((1-\gamma_k) + \sum_{j=1}^{n} \frac{(b_k^j)^2}{2c_k^{(a)j}} A_{k+1}^j \delta\right)\delta,$$

$$B_k^i = \frac{(\varpi_k^i)^2}{\varepsilon_k^i} + B_{k+1}^i(1-\lambda_k)\delta \quad \text{for } i \in N, \quad \text{and}$$
$$k \in \{T-2, T-3, \ldots, 1\}, \tag{24}$$

and C_k^i *is a function involving game equilibrium parameters.*

(2) *The game equilibrium strategies are*

$$\bar{q}_k^{i**} = \frac{\alpha_k^i - c_k^i + A_{k+1}^i \delta a_k^i}{2\beta_k^i}, \quad q_k^{i**} = \frac{(\sigma_k^i)^2 \vartheta_k}{(\varepsilon_k^i)^2},$$
$$u_k^{i**} = -\frac{b_k^i}{2c_k^{(a)i}} A_{k+1}^i \delta (x_k)^{1/2}, \quad w_k^{i**} = \frac{\chi_k^i + B_{k+1}^i \delta z_k^i}{2} \tag{25}$$

for $i \in N$, *and* $k \in \{T-2, T-3, \ldots, 1\}$.

Proof. See Appendix A. □

6.2. Group Optimal Outcome

Now that the nations agree to cooperate, the joint payoffs of the nations under collaboration and the optimal cooperative strategies are given in the following proposition.

Proposition 6.2. (1) *The joint payoff of the nations under collaboration is*

$$W^N(k, x, \vartheta) = (A_k x_k + B_k \vartheta_k + C_k)\,\delta^{k-1}, \tag{26}$$

where

$$A_k = \sum_{i=1}^n \left[-c_k^{(a)i} \left(\frac{b_k^i}{2c_k^{(a)i}} A_{k+1}\delta \right)^2 - h_k^i \right] + A_{k+1}\left((1-\gamma_k) + \sum_{j=1}^n \frac{(b_k^j)^2}{2c_k^{(a)j}} A_{k+1}\delta \right)\delta, \tag{27}$$

$$B_k = \sum_{i=1}^n \frac{(\varpi_k^i)^2}{\varepsilon_k^i} + B_{k+1}(1-\lambda_k)\delta, \tag{28}$$

and C_k *is a function of the model parameters.*

(2) *The optimal controls under collaboration are*

$$\bar{q}_k^{i*} = \frac{\alpha_k^i - c_k^i + A_{k+1}\delta a_k^i}{2\beta_k^i}, \quad q_k^{i*} = \frac{(\varpi_k^i)^2\vartheta_k}{(\varepsilon_k^i)^2},$$

$$u_k^{i*} = -\frac{b_k^i}{2c_k^{(a)i}}A_{k+1}\delta\,(x_k)^{1/2}, \quad \bar{w}_k^{i*} = \frac{\chi_k^i + B_k\delta z_{k+1}^i + B_{k+2}\delta^2 z_k^{(1)i}}{2} \tag{29}$$

for $i \in N$ *and* $k \in \{T-2, T-3, \ldots, 1\}$.

Proof. See Appendix B. □

Substituting the relevant optimal strategies in (29) into the state dynamics (21) and (22), we obtain the optimal cooperative dynamics as

$$x_{k+1} = x_k + \sum_{i=1}^{n} a_k^i \bar{q}_k^{i*} - \sum_{i=1}^{n} b_k^i u_k^{i*}\,(x_k)^{1/2} - \gamma_k x_k, x_1 = x_1^0,$$

$$\vartheta_{k+1} = \vartheta_k + \sum_{i=1}^{n} z_k^i w_k^{i*} - \lambda_k\vartheta_k, \quad \vartheta_1 = \vartheta_1^0. \tag{30}$$

We use $\{x_k^*, \vartheta_k^*\}_{k=1}^{T+1}$ to denote the solution to system (30).

6.3. Fair Gain-Sharing Weight

To obtain the fair gain-sharing weight, we first have to obtain $W^{(N\backslash i)}(k, x, \vartheta)$ by solving the game:

$$\max_{\substack{q_k^\ell, \bar{q}_k^\ell, w_k^\ell, u_k^\ell \\ \ell \in N\backslash i \\ k\in\{1,2,\ldots,T\}}} \left\{ \sum_{\ell\in N\backslash i} \left[\sum_{k=1} (\alpha_k^\ell \bar{q}_k^\ell - \beta_k^\ell (\bar{q}_k^\ell)^2 - c_k^\ell \bar{q}_k^\ell + 2\varpi_k^\ell(\vartheta_k q_k^\ell)^{1/2} - \varepsilon_k^\ell q_k^\ell \right.\right.$$

$$+ \chi_k^\ell w_k^\ell - (w_k^\ell)^2 - c_k^{(a)\ell}(u_k^\ell)^2 - h_k^\ell x_k)\delta^{k-1}$$

$$\left.\left. +(h_{T+1}^\ell x_{T+1} + y_{T+1}^\ell \vartheta_{T+1} + Q_{T+1}^\ell)\delta^T \right] \right\}, \tag{31}$$

$$\max_{\substack{q_k^i,\bar{q}_k^i,w_k^i,u_k^i\\ k\in\{1,2,\ldots,T\}}} \left\{ \sum_{k=1}^{T} [\alpha_k^i \bar{q}_k^i - \beta_k^i (\bar{q}_k^i)^2 - c_k^i \bar{q}_k^i + 2\varpi_k^i (\vartheta_k q_k^i)^{1/2} - \varepsilon_k^i q_k^i \right.$$

$$+ \chi_k w_k^i - (w_k^i)^2 - c_k^{(a)i} (u_k^i)^2 - h_k^i x_k] \delta^{k-1}$$

$$\left. + (h_{T+1}^i x_{T+1} + y_{T+1}^i \vartheta_{T+1} + Q_{T+1}^i) \delta^T \right\} \tag{32}$$

subject to dynamics (21).

The system (31)-(32) and (21) form a two-player game. The solution to the game can be characterized by the following theorem.

Theorem 6.1. *A set of strategies* $\{(\hat{q}_k^i, \hat{\bar{q}}_k^i, \widehat{w}_k^i, \widehat{u}_k^i)$ *and* $(q_k^{(N\backslash i)\ell}, \bar{q}_k^{(N\backslash i)\ell}, w_k^{(N\backslash i)\ell}, u_k^{(N\backslash i)\ell})$ *for* $\ell \in N\backslash i\}$ *constitutes a set of feedback Nash equilibria of the game (31)-(32) and (21)) if there exist functions* $W^{N\backslash i}(k,x)$ *and* $\hat{V}^i(k,x)$, *which satisfy the following recursive equations:*

$$W^{N\backslash i}(T+1, x, \vartheta) = \sum_{j\in N\backslash i} (h_{T+1}^\ell x_{T+1} + y_{T+1}^\ell \vartheta_{T+1} + Q_{T+1}^\ell) \delta^T \tag{33}$$

$$W^{N\backslash i}(k, x, \vartheta) = \max_{\substack{q_k^\ell,\bar{q}_k^\ell,w_k^\ell,u_k^\ell\\ \ell\in N\backslash i}} \left\{ \sum_{\ell\in N\backslash i}^{\ell} (\alpha_k^\ell \bar{q}_k^\ell - \beta_k^\ell (\bar{q}_k^\ell)^2 - c_k^\ell \bar{q}_k^\ell \right.$$

$$+ 2\varpi_k^\ell (\vartheta_k q_k^\ell)^{1/2} - \varepsilon_k^\ell q_k^\ell + \chi_k^\ell w_k^\ell - (w_k^\ell)^2$$

$$- c_k^{(a)\ell} (u_k^\ell)^2 - h_k^\ell x_k) \delta^{k-1}$$

$$\left. + W^{N\backslash i}(k+1, x_{k+1}^{N\backslash i}, \vartheta_{k+1}^{N\backslash i}) \right\} \tag{34}$$

for $k \in \{1, 2, \ldots, T\}$,
where

$$x_{k+1}^{N\backslash i} = x_k + \sum_{\ell\in N\backslash i} a_k^\ell \bar{q}_k^\ell + a_k^i \hat{\bar{q}}_k^i - \sum_{\ell\in N\backslash i} b_j u_k^\ell (x_k)^{1/2}$$

$$- b_k^i \hat{u}_k^i (x_k)^{1/2} - \gamma_k x_k,$$

$$\vartheta_{k+1}^{N\backslash i} = \vartheta_k + \sum_{\ell \in N\backslash i} z_k^\ell w_k^\ell + z_k^i \widehat{w}_k^i - \lambda_k \vartheta_k;$$

$$\widehat{V}^i(T+1, x, \vartheta) = (h_{T+1}^i x_{T+1} + y_{T+1}^i \vartheta_{T+1} + Q_{T+1}^i)\delta^T \tag{35}$$

$$\begin{aligned}\widehat{V}^i(T+1, x, \vartheta) = \max_{q_k^i, \bar{q}_k^i, \bar{w}_k^i, u_k^i} & \Big\{(\alpha_k^i \bar{q}_k^i - \beta_k^i (\bar{q}_k^i)^2 - c_k^i \bar{q}_k^i \\ & + 2\varpi_k^i (\vartheta_k q_k^i)^{1/2} - \varepsilon_k^i q_k^i \\ & + \chi_k w_k^i - (w_k^i)^2 - c_k^{(a)i} (u_k^i)^2 - h_k^i x_k)\delta^{k-1} \\ & + \hat{V}^i(k+1, x_{k+1}^{(i)i} \vartheta_{k+1}^{(i)i})\Big\}\end{aligned} \tag{36}$$

for $k \in \{1, 2, \ldots, T\}$,

where

$$\begin{aligned} x_{k+1}^{(i)i} = & \, x_k + a_k^i \bar{q}_k^i + \sum_{\ell \in N\backslash i} a_k^\ell \bar{q}_k^{(N\backslash i)\ell} - b_k^i u_k^i (x_k)^{1/2} \\ & - \sum_{\ell \in N\backslash i} b_j u_k^{(n\backslash i)\ell} (x_k)^{1/2} - \gamma_k x_k, \end{aligned}$$

$$\vartheta_{k+1}^{(i)i} = \vartheta_k + z_k^i w_k^i + \sum_{\ell \in N\backslash i} z_k^\ell w_k^{(N\backslash i)\ell} - \lambda_k \vartheta_k.$$

Performing the indicated maximization operators in Theorem 6.1, we obtain the game equilibrium payoffs and strategies of the nations in the following proposition.

Proposition 6.3. (1) *The game equilibrium payoff of the nations in* $N\backslash i$ *under collaboration is*

$$W^{N\backslash i}(k, x, \vartheta) = (A_k^{N\backslash i} x_k + B_k^{N\backslash i} \vartheta_k + C_k^{N\backslash i})\delta^{k-1},$$

where

$$\begin{aligned} A_k^{N\backslash i} = & \sum_{\ell \in N\backslash i} \left[-c_k^{(a)\ell} \left(\frac{b_k^\ell}{2c_k^{(a)\ell}} A_{k+1}^{N\backslash i} \delta \right)^2 - h_k^\ell \right] \\ & + A_{k+1}^{N\backslash i} \left((1 - \gamma_k) + \sum_{\ell \in N\backslash i} \frac{(b_k^\ell)^2}{2c_k^{(a)\ell}} A_{k+1}^{N\backslash i} \delta + \frac{(b_k^i)^2}{2c_k^{(a)i}} \hat{A}_{k+1}^i \delta \right) \delta, \end{aligned}$$

$$B_k^{N\backslash i} = \sum_{\ell \in N\backslash i} \frac{\left(\varpi_k^\ell\right)^2}{\varepsilon_k^\ell} + B_{k+1}^{N\backslash i}\left(1-\lambda_k\right)\delta,$$

and $C_k^{N\backslash i}$ *is a function involving the model parameters.*
(2) *The game equilibrium strategies of coalition* $N\backslash i$ *are*

$$\bar{q}_k^{(N\backslash i)\ell} = \frac{\alpha_k^\ell - c_k^\ell + A_{k+1}^{N\backslash i}\delta a_k^\ell}{2\beta_k^\ell}, \quad q_k^{(N\backslash i)\ell} = \frac{\left(\varpi_k^\ell\right)^2 \vartheta_k}{\left(\varepsilon_k^\ell\right)^2},$$

$$u_k^{(N\backslash i)\ell} = -\frac{b_k^\ell}{2c_k^{(a)\ell}} A_{k+1}^{N\backslash i}\delta\left(x_k\right)^{1/2}, \quad w_k^{(N\backslash i)\ell} = \frac{\chi_k^\ell + B_{k+1}^{N\backslash i}\delta z_k^\ell}{2} \tag{37}$$

for $\ell \in N\backslash i$ *and* $k \in \{T-2, T-3, \ldots, 1\}$.
(3) *The game equilibrium payoff of the nation* i *is*

$$\hat{V}^i(k, x, \vartheta) = (\hat{A}_k^i x + \hat{B}_k^i \vartheta + \hat{C}_k^i)\delta^{k-1}, \tag{38}$$

where

$$\begin{aligned}
\hat{A}_k^i &= -c_k^{(a)i}\left(\frac{b_k^i}{2c_k^{(a)i}}\hat{A}_{k+1}^{N\backslash i}\delta\right)^2 - h_k^i \\
&\quad + \hat{A}_{k+1}^i\left(\left(1-\gamma_k\right) + \sum_{\ell \in N\backslash i}\frac{\left(b_k^\ell\right)^2}{2c_k^{(a)\ell}}A_{k+1}^{(N\backslash i)}\delta + \frac{\left(b_k^i\right)^2}{2c_k^{(a)i}}\hat{A}_{k+1}^i\delta\right)\delta
\end{aligned}$$

$$\hat{B}_k^i = \frac{\left(\omega_k^i\right)^2}{\varepsilon_k^i} + \hat{B}_{k+1}^i\left(1-\lambda_k\right)\delta$$

and $\hat{C}_k^i$ *is a function of the model parameters.*
(4) *The game equilibrium strategies of nation* i *are*

$$\hat{\bar{q}}_k^i = \frac{\alpha_k^i - c_k^i + \hat{A}_{k+1}^i\delta a_k^i}{2\beta_k^i}, \quad \hat{q}_k^i = \frac{\left(\varpi_k^i\right)^2\vartheta_k}{\left(\varepsilon_k^i\right)^2},$$

$$\widehat{u}_k^i = -\frac{b_k^i}{2c_k^{(a)i}}\hat{A}_{k+1}^i\delta\left(x_k\right)^{1/2}, \quad \widehat{w}_k^i = \frac{\chi_k^i + \hat{B}_k^i\delta z_{k+1}^i + \hat{B}_{k+2}^i\delta^2 z_k^{(1)i}}{2} \tag{39}$$

for $k \in \{T-2, T-3, \ldots, 1\}$.

Proof. Follow the proofs in Appendices A and B . A fair gain-sharing weight can be obtained as

$$\frac{W^N(k,x,\vartheta) - W^{N\backslash i}(k,x,\vartheta)}{\sum_{\zeta=1}^{n}[W^N(k,x,\vartheta) - W^{N\backslash\varsigma}(k,x,\vartheta)]}$$
$$= \frac{A_k x_k + B_k\vartheta_k + C_k - (A_k^{N\backslash i}x_k + B_k^{N\backslash i}\vartheta_k + C_k^{N\backslash i})}{\sum_{\varsigma=1}^{n}[A_k x_k + B_k\vartheta_k + C_k - (A_k^{N\backslash\varsigma}x_k + B_k^{N\backslash\varsigma}\vartheta_k + C_k^{N\backslash\varsigma})]} \tag{40}$$

for $i \in N$. □

6.4. A Dynamically Stable Fair Gain-Sharing Value

Using Formula 4.1, a dynamically stable fair gain-sharing value can be obtained as

$$\xi^i(k,x,\vartheta) = \frac{W^N(k,x,\vartheta) - W^{N\backslash i}(k,x,\vartheta)}{\sum_{\varsigma=1}^{n}\left[W^N(k,x,\vartheta) - W^{N\backslash\varsigma}(k,x,\vartheta)\right]}$$
$$\times \left[W^N(k,x,\vartheta) - \sum_{j=1}^{n} V^j(k,x,\vartheta)\right] + V^i(k,x,\vartheta)$$
$$= \frac{A_k x_k + B_k\vartheta_k + C_k - (A_k^{N\backslash i}x_k + B_k^{N\backslash i}\vartheta_k + C_k^{N\backslash i})}{\sum_{\varsigma=1}^{n}[A_k x_k + B_k\vartheta_k + C_k - (A_k^{N\backslash\varsigma}x_k + B_k^{N\backslash\varsigma}\vartheta_k + C_k^{N\backslash\varsigma})]}$$
$$\times \left(A_k x_k + B_k\vartheta_k + C_k - \sum_{j=1}^{n}(A_k^j x + B_k^j\vartheta + C_k^j)\right)\delta^{k-1}$$
$$+ \left(A_k^i x + B_k^i\vartheta + C_k^i\right)\delta^{k-1} \tag{41}$$

for $i \in N$ along the cooperative state path $\{x_k^*, \vartheta_k^*\}_{k=1}^{T}$.

7. Concluding Remarks

Given the asymmetry of the nations in a cooperating scheme, some nations may capture substantially less gain than their contributions to other nations. A fair gain-sharing principle of cooperative gain is critical to obtain the agreement of all nations to cooperate and ensure the scheme's sustainability. In this chapter, we first provide a fair gain-sharing cooperation scheme, which satisfies (i) individual rationality, (ii) group optimality, (iii) fair gain-sharing property that reflects the contributions of nations to the cooperative gains, and (iv) time consistency that guarantees that the agreed-upon gain-sharing principle is maintained in every stage of the cooperative duration.

A comprehensive review and a future research agenda for gain-sharing in business were offered by Welbourne and Mejia (1995). Mangel and Useem (2000) explored the strategic significance of gain-sharing within the labor market. Mohamed et al. (2020) introduced a knowledge-based algorithm for gain-sharing aimed at addressing optimization challenges. O'Byrne and Rajgopal (2022) evaluated employee value added through a novel measure of gain-sharing between labor and capital. Carlan et al. (2022) investigated cost-effectiveness and gain-sharing scenarios related to the acquisition of a blockchain-based application in the maritime supply chain.

The dynamically stable fair gain-sharing value presented in (18) was formulated by Yeung et al. (2024a). A fair-sharing value for cooperative dynamic network games was defined by Yeung et al. (2024b). Yeung et al. (2024c) introduced a dynamic collaborative solution featuring a fair gain-sharing property for the management of the interconnection of electricity generation, transmission, and distribution.

Appendices

Appendix A. Proof of Proposition 6.1

Invoking Theorem 2.1, a feedback Nash equilibrium solution of the game (21)–(23) can be characterized by the following corollary.

Corollary A.1. *A set of strategies $\{q_k^{i**}, \bar{q}_k^{i**}, u_k^{i**}, w_k^{i**}$ for $k \in \{1, 2, \ldots, T\}$ and $i \in N\}$ provides a feedback Nash equilibrium solution of the game (21)–(23) if there exist functions $V^i(k, x, \vartheta)$ such that the following recursive relations are satisfied:*

$$\begin{aligned} V^i(T+1, x, \vartheta) &= \left(A_{T+1}^i x_{T+1} + B_{T+1}^i \vartheta_{T+1} + C_{T+1}^i\right) \delta^T \\ &= (-h_{T+1}^i x_{T+1} + y_{T+1}^i \vartheta_{T+1} + Q_{T+1}^i)\delta^T \end{aligned} \tag{A.1}$$

$$
\begin{aligned}
V^i(k,x,\vartheta) = \max_{q_k^i,\bar{q}_k^i,w_k^i,u_k^t} & \Big\{[\alpha_k^i\bar{q}_k^i - \beta_k^i(\bar{q}_k^i)^2 - c_k^i\bar{q}_k^i + 2\varpi_k(\vartheta_k q_k^i)^{1/2} \\
& - \varepsilon_k^i q_k^i + \chi_k w_k^i - (w_k^i)^2 - c_k^{(a)i}(u_k^i)^2 - h_k^i x_k]\delta^{k-1} \\
& + V^i(k+1, x_{k+1}^{\pm i*}, \vartheta_{k+1}^{\pm i**})\Big\} \qquad \text{(A.2)}
\end{aligned}
$$

for $k \in \{1, 2, \ldots, T\}$ and $i \in N$.

where

$$
\begin{aligned}
x_{k+1}^{\neq i**} &= x_k + a_k^i\bar{q}_k^i + \sum_{\substack{j=1 \\ i\neq j}}^{n} a_k^j\bar{q}_k^{j**} - b_k^i u_k^i\,(x_k)^{1/2} \\
&\quad - \sum_{\substack{j=1 \\ j\neq i}}^{n} b_j u_k^{j**}\,(x_k)^{1/2} - \gamma_k x_k, \\
\vartheta_{k+1}^{\neq i**} &= \vartheta_k + z_k^i w_k^i + \sum_{\substack{j=1 \\ j\neq i}}^{n} z_k^j w_k^{j**} - \lambda_k\vartheta_k.
\end{aligned}
$$

Using (A.1), *we obtain*

$$
A_{T+1}^i = -h_{T+1}^i, B_{T+1}^i = y_{T+1}^i, C_{T+1}^i = Q_{T+1}^i. \qquad \text{(A.3)}
$$

Invoking the technique of backward induction, we consider first the last period of operation, i.e., period T. Using (A.2) *and* (A.3), *we can obtain the game equilibrium strategies in period T as*

$$
\begin{aligned}
\bar{q}_T^i &= \frac{\alpha_T^i - c_T^i + A_{T+1}^i\delta a_T^i}{2\beta_T^i}, \quad q_T^i = \frac{(\varpi_T^i)^2\vartheta_T}{(\varepsilon_T^i)^2}, \\
u_T^i &= -\frac{b_T^i}{2c_T^{(a)i}}A_{T+1}^i\delta(x_T)^{1/2}, w_T^i = \frac{\chi_T^i + B_{T+1}^i\delta z_T^i}{2} \qquad \text{(A.4)}
\end{aligned}
$$

for $i \in N$.

Substituting the game equilibrium strategies in (A.4) *into the period* T *equation of* (A.2), *we obtain*

$$V^i(T,x,\vartheta) = \left[\alpha_T^i \frac{\alpha_T^i - c_T^i + A_{T+1}^i \delta a_T^i}{2\beta_T^i} - \beta_T^i \left(\frac{\alpha_T^i - c_T^i + A_{T+1}^i \delta a_T^i}{2\beta_T^i}\right)^2 \right.$$
$$- c_T^i \frac{\alpha_T^i - c_T^i + A_{T+1}^i \delta a_T^i}{2\beta_T^i} + \frac{\left(\varpi_T^i\right)^2 \vartheta_T}{\varepsilon_T^i} + \chi_T \frac{\chi_T^i + B_{T+1}^i \delta z_T^i}{2}$$
$$- \left(\frac{\chi_T^i + B_{T+1}^i \delta z_T^i}{2}\right)^2$$
$$\left. - c_T^{(a)i} \left(\frac{b_T^i}{2c_T^{(a)i}} A_{T+1}^i \delta \left(x_T\right)^{1/2}\right)^2 - h_T^i x_T \right] \delta^{T-1}$$
$$+ \left[A_{T+1}^i \left(x_T + \sum_{j=1}^n a_T^j \frac{\alpha_T^j - c_T^j + A_{T+1}^j \delta a_T^j}{2\beta_T^i}\right.\right.$$
$$\left. + \sum_{j=1}^n \frac{(b_T^j)^2}{2c_T^{(a)j}} A_{T+1}^j \delta x_T - \gamma_T x_T \right)$$
$$\left. + B_{T+1}^i \left(\vartheta_T + \sum_{j=1}^n z_T^j \frac{\chi_T^j + B_{T+1}^j \delta z_T^j}{2} - \lambda_T \vartheta_T\right) + C_{T+1}^i \right] \delta^T \tag{A.5}$$

for $i \in N$.

The RHS of the equation system (A.5) *is a linear function of* x *and* ϑ *which can be expressed as* $(A_T^i x_T + B_T^i \vartheta_T + C_T^i)\delta^{T-1}$, *where*

$$A_T^i x_T = -c_T^{(a)i} \left(\frac{b_T^i}{2c_T^{(a)i}} A_{T+1}^i \delta\right)^2 x_T - h_T^i x_T$$
$$+ A_{T+1}^i \left(\quad (1-\gamma_k)\, x_T + \sum_{j=1}^n \frac{\left(b_T^j\right)^2}{2c_T^{(a)i}} A_{T+1}^i \delta x_T \right) \delta,$$

$$B_T^i \vartheta_T = \frac{\left(\varpi_T^i\right)^2 \vartheta_T}{\varepsilon_T^i} + B_{T+1}^i \left(1 - \lambda_T\right) \vartheta_T \delta,$$

and C_T^i is a function involving the terms in (A.5) that do not involve x_T and ϑ_T.

We consider the processes for periods $T-1, T-2, \ldots, 1$. We can obtain the game equilibrium strategies in period $k \in \{T-1, T-2, \ldots, 1\}$ as

$$\bar{q}_k^i = \frac{\alpha_k^i - c_k^i + A_{k+1}^i \delta a_k^i}{2\beta_k^i}, \quad q_k^i = \frac{\left(\varpi_k^i\right)^2 \vartheta_k}{\left(\varepsilon_k^i\right)^2},$$

$$u_k^i = -\frac{b_k^i}{2c_k^{(a)i}} A_{k+1}^i \delta \left(x_k\right)^{1/2}, \quad w_k^i = \frac{\chi_k^i + B_{k+1}^i \delta z_k^i}{2} \tag{A.6}$$

for $i \in N$.

Substituting the game equilibrium strategies in (A.6) into the period k equation of (A.2), we obtain an equation system whose RHS is a linear function of x and ϑ, which can be expressed as $(A_k^i x_k + B_k^i \vartheta_k + C_k^i)\delta^{k-1}$, where

$$A_k^i = -c_k^{(a)i} \left(\frac{b_k^i}{2c_k^{(a)i}} A_{k+1}\delta\right)^2 - h_k^i$$

$$+ A_{k+1}^i \left((1 - \gamma_k) + \sum_{j=1}^{n} \frac{\left(b_k^j\right)^2}{2c_k^{(a)j}} A_{k+1}^j \delta \right) \delta,$$

$$B_k^i = \frac{\left(\varpi_k^i\right)^2}{\varepsilon_k^i} + B_{k+1}^i \left(1 - \lambda_k\right) \delta, \tag{A.7}$$

and C_k^i is a function involving the model parameters for $i \in N$, and $k \in \{T-2, T-3, \ldots, 1\}$.

Appendix B. Proof of Proposition 6.2

Invoking Theorem 2.2, the cooperative solution of the game (21)–(23) can be characterized by the following corollary.

Corollary B.1. *A set of controls* $\{q_k^{i*}, \bar{q}_k^{i*}, u_k^{i*}, w_k^{i*}$ *for* $k \in \{1, 2, \ldots, T\}$ *and* $i \in N\}$ *provides an optimal collaborative solution to the game (21)-(23) if there exist functions* $W^N(k, x, \vartheta)$ *such that the following recursive relations are satisfied:*

$$\begin{aligned} W^N(T+1, x, \vartheta) &= \sum_{i=1}^{n} \left(A_{T+1}x_{T+1} + B_{T+1}\vartheta_{T+1} + C_{T+1}\right)\delta^T \\ &= \sum_{i=1}^{n} \left(-h_{T+1}^i x_{T+1} + y_{T+1}^i \vartheta_{T+1} + Q_{T+1}^i\right)\delta^T \end{aligned} \tag{B.1}$$

$$\begin{aligned} W^N(k, x, \vartheta) = \max_{q_k^i, \bar{q}_k^i, w_k^i, u_k^t} &\left\{ \sum_{i=1}^{n} [\alpha_k^i \bar{q}_k^i - \beta_k^i \left(\bar{q}_k^i\right)^2 - c_k^i \bar{q}_k^i \right. \\ &+ 2\varpi_k(\vartheta_k q_k^i)^{1/2} - \varepsilon_k^i q_k^i + \chi_k w_k^i - (w_k^i)^2 - c_k^{(a)i} \\ &\left. \times (u_k^i)^2 - h_k^i x_k]\delta^{k-1} + V^i\left(k+1, x_{k+1}, \vartheta_{k+1}\right) \right\} \end{aligned} \tag{B.2}$$

for $k \in \{1, 2, \ldots, T\}$ *and* $i \in N$,
where

$$\begin{aligned} x_{k+1} &= x_k + \sum_{j=1}^{n} a_k^j \bar{q}_k^j - \sum_{j=1}^{n} b_j u_k^j \left(x_k\right)^{1/2} - \gamma_k x_k, \\ \vartheta_{k+1} &= \vartheta_k + \sum_{j=1}^{n} z_k^j w_k^j - \lambda_k \vartheta_k. \end{aligned}$$

Using (B.1), we obtain

$$A_{T+1} = \sum_{i=1}^{n} -h_{T+1}^i, B_{T+1} = \sum_{i=1}^{n} y_{T+1}^i, C_{T+1} = \sum_{i=1}^{n} Q_{T+1}^i. \tag{B.3}$$

Using the technique of backward induction, we consider first the last period of operation, i.e., period T. *Using (B.2) and (B.3), we* *can obtain the optimal collaborative controls in period T as*

$$\bar{q}_T^{i*} = \frac{\alpha_T^i - c_T^i + A_{T+1}\delta a_T^i}{2\beta_T^i}, q_T^{i*} = \frac{\left(\omega_T^i\right)^2 \vartheta_T}{\left(\varepsilon_T^i\right)^2},$$

$$u_T^{i*} = -\frac{b_T^i}{2c_T^{(a)i}} A_{T+1}\delta(x_T)^{1/2}, w_T^{i*} = \frac{\chi_T^i + B_{T+1}\delta z_T^i}{2} \quad \text{for } i \in N. \tag{B.4}$$

Substituting the optimal controls in (B.4) into the period T equation of (B.2), we obtain

$$\begin{aligned}
W(T, x, \vartheta) = \sum_{i=1}^{n} \Bigg[& \alpha_T^i \frac{\alpha_T^i - c_T^i + A_{T+1}^i \delta a_T^i}{2\beta_T^i} \\
& - \beta_T^i \left(\frac{\alpha_T^i - c_T^i + A_{T+1}^i \delta a_T^i}{2\beta_T^i} \right)^2 \\
& - c_T^i \frac{\alpha_T^i - c_T^i + A_{T+1}^i \delta a_T^i}{2\beta_T^i} + \frac{\left(\omega_T^i\right)^2 \vartheta_T}{\varepsilon_T^i} \\
& + \chi_T \frac{\chi_T^i + B_{T+1}^i \delta z_T^i}{2} - \left(\frac{\chi_T^i + B_{T+1}^i \delta z_T^i}{2} \right)^2 \\
& - c_T^{(a)i} \left(\frac{b_T^i}{2c_T^{(a)i}} A_{T+1}^i \delta \left(x_T\right)^{1/2} \right)^2 - h_T^i x_T \Bigg] \delta^{T-1} \\
& + \Bigg[A_{T+1} \left(x_T \sum_{j=1}^{n} \alpha_T^j \frac{\alpha_T^j - c_T^j + A_{T+1}\delta a_T^j}{2\beta_T^i} \right. \\
& \left. + \sum_{j=1}^{n} \frac{(b_T^j)^2}{2c_T^{(a)j}} A_{T+1}\delta x_T - \gamma_T x_T \right) \\
& + B_{T+1} \left(\vartheta_T + \sum_{j=1}^{n} z_T^j \frac{\chi_T^j + B_{T+1}^j \delta z_T^j}{2} - \lambda_T \vartheta_T \right) \\
& + C_{T+1}^i \Bigg] \delta^T
\end{aligned} \tag{B.5}$$

The RHS of the equation system (B.5) is a linear function of x *and* ϑ, *which can be expressed as* $(A_T x_T + B_T \vartheta_T + C_T)\delta^{T-1}$, *where*

$$A_T x_T = \sum_{i=1}^{n} \left[-c_T^{(a)i} \left(\frac{b_T^i}{2c_T^{(a)i}} A_{T+1}\delta \right)^2 - h_t^i x_T \right]$$

$$+ A_{T+1} \left((1-\gamma_T) x_T + \sum_{j=1}^{n} \frac{\left(b_T^j\right)^2}{2c_T^{(a)j}} A_{T+1} \delta x_T \right) \delta$$

$$B_T \vartheta_T = \sum_{i=1}^{n} \frac{\left(\omega_T^i\right)^2 \vartheta_T}{\varepsilon_T^i} + B_{T+1}(1-\lambda_T)\vartheta_T \delta;$$

and C_T *is a function involving the terms in (B.5) that do not involve* x_T *and* ϑ_T.

Repeating the processes for periods $T-1, T-2, \ldots, 1$, *we can obtain the optimal controls in period* $k \in \{T-1, T-2, \ldots, 1\}$ *as*

$$\bar{q}_k^{i*} = \frac{\alpha_k^i - c_k^i + A_{k+1}\delta a_k^i}{2\beta_k^i}, \quad q_k^{i*} = \frac{\left(\sigma_k^i\right)^2 \vartheta_k}{\left(\varepsilon_k^i\right)^2},$$

$$u_k^{i*} = -\frac{b_k^i}{2c_k^{(a)i}} A_{k+1}\delta (x_k)^{1/2}, \quad \bar{w}_k^{i*} = \frac{\chi_k^i + B_{k+1}\delta z_k^i}{2} \tag{B.6}$$

for $i \in N$.

Substituting the optimal strategies in (B.6) into the period k *equation of (B.2), we obtain an equation system whose RHS is a linear function of* x *and* ϑ, *which can be expressed as* $(A_k x_k + B_k \vartheta_k + C_k)\delta^{k-1}$, *where*

$$A_k = \sum_{i=1}^{n} \left[-c_k^{(a)i} \left(\frac{b_k^i}{2c_k^{(a)i}} A_{k+1}\delta \right)^2 - h_k^i \right]$$

$$+ A_{k+1} \left((1-\gamma_k) + \sum_{j=1}^{n} \frac{\left(b_k^j\right)^2}{2c_k^{(a)j}} A_{k+1}\delta \right) \delta,$$

$$B_k = \sum_{i=1}^{n} \frac{\left(\varpi_k^i\right)^2}{\varepsilon_k^i} + B_{k+1}(1-\lambda_k)\delta, \tag{B.7}$$

and C_k *is a function involving the model parameters for* $i \in N$ *and* $k \in \{T-1, T-2, \ldots, 1\}$.

References

Carlan, V., Sys, C., & Vanelslander, T. (2022). Cost-effectiveness and gain-sharing scenarios for purchasing a blockchain-based application in the maritime supply chain. *European Transport Research Review*, 14, 21. https://doi.org/10.1186/s12544-022-00545-2.

Mangel, R., & Useem, M. (2000). The strategic role of gainsharing. *Journal of Labor Research*, 21, 327–343. https://doi.org/10.1007/s12122-000-1051-y.

Mohamed, A. W., Hadi, A. A., & Mohamed, A. K. (2020). Gaining-sharing knowledge based algorithm for solving optimization problems: A novel nature-inspired algorithm. *International Journal of Machine Learning and Cybernetics*, 11(7), 1501–1529. https://doi.org/10.1007/s13042-019-01053-x.

O'Byrne, S. F., & Rajgopal, S. (2022). Employee value added: A new measure of gainsharing between labor and capital. *Journal of Applied Corporate Finance*, 34(2), 30–44. https://doi.org/10.1111/jacf.12503.

Welbourne, T. M., & Mejia, L. R. G. (1995). Gainsharing: A critical review and a future research agenda. *Journal of Management*, 21(3), 559–609. https://doi.org/10.1177/014920639502100307.

Yeung, D. W. K., Petrosyan, L. A., & Zhang, Y. X. (2024a). Managing interconnected production in a dynamic network: A fair gain-sharing collaborative solution. *Contributions to Game Theory and Management*, 17, 231–242.

Yeung, D. W. K., Petrosyan, L. A., & Zhang, Y. X. (2024b). A fair-sharing value for cooperative dynamic network games. *Dynamic Games and Applications*. https://doi.org/10.1007/s13235-024-00562-x.

Yeung, D. W., Petrosyan, L. A., & Zhang, Y. (2024c). Managing interconnection of electricity generation, transmission and distribution: A dynamic collaborative solution. In *TENCON 2024–2024 IEEE Region 10 Conference (TENCON)* (pp. 994–997). IEEE.

Chapter 4

Dynamic Joint Ventures and Fair Gain-Sharing

"When entering into a partnership, it's crucial to set the stage for success by safeguarding your business's interests. This requires proactive planning to anticipate and address potential risks and challenges that may arise...... Profit-sharing agreements act as the compass that steers your partnership toward prosperity and business growth."

— **Joy Cunanan**

Joint ventures (JVs) have led to unprecedented and rapid growth in the global economy, driven by globalization, over the last 30 years. JVs are prevalent in almost all developed and developing economies. Nevertheless, JVs are encountering crucial challenges, specifically, the sustainability of collaboration, the existence of acceptable gain-sharing arrangements, and the consistency of the original agreement over time. These unresolved issues have resulted in serious consequences, such as the disruptive termination of private–private ventures, international legal disputes involving government–private ventures, and uneventful government–government ventures. This chapter formulates a cooperative solution for JVs that includes (i) individual and group rationalities, (ii) a fair gain-sharing arrangement, and (iii) the consistency of the agreed-upon terms. Section 1 examines joint ventures and globalization. Section 2 introduces a general game framework for joint ventures, the benchmark non-cooperative equilibria, the payoff structure under the joint venture, and the maximized venture payoff. Section 3 addresses gain-sharing in a dynamic joint venture and the marginal contributions of venture participants. Section 4 offers a dynamic fair-sharing cooperative solution for JVs. Section 5 provides an illustrative example, and Section 6 provides the concluding remarks.

1. Joint Ventures and Globalization

Globalization and JVs have resulted in an exceptionally rapid expansion of the global economy over the last 30 years. JVs are crucial to globalization as they promote collaboration both across borders and within domestic markets. These ventures are present in nearly all developed and developing nations, encompassing private–private, government–private, and government–government partnerships. By leveraging the intricate nature of the international division of labour through the offshoring of production activities and the transfer of technology, JVs significantly enhance the benefits derived from trade. This innovative production organization is not limited to large corporations; rather, it is driving changes that affect the entire value chain across virtually all sectors and types of businesses. The relationship between JVs and globalization is deeply interconnected, with substantial consequences for the advancement of the global economy.

The contributions of JVs include the following:

- **Access to New Markets:** JVs enable companies to penetrate foreign markets more effectively by utilizing the expertise and networks of local partners.
- **Sharing Expertise and Resources:** Organizations can combine their strengths, including technology, infrastructure, and human capital, to improve efficiency and foster innovation.
- **Risk Mitigation:** By distributing financial and operational risks, companies can lessen the uncertainties linked to international expansion.
- **Enhanced Innovation and Competitiveness:** The partnership between firms from various regions promotes creativity and technological progress.

- **Cultural and Market Adaptability:** JVs assist businesses in adjusting to local consumer preferences and cultural subtleties, thereby enhancing the appeal of their products and services.
- **Technology Spillover:** Companies engaged in JVs benefit from technology spillover as they collaboratively pursue their activities.

Instances of JVs are prevalent in the contemporary business world. For instance, China's international initiative known as the Belt and Road Initiative (BRI) facilitates the formation of international JVs and seeks to allocate substantial capital across various participating nations. As reported by Daxue Consulting Beijing, there were 124,600 Sino-foreign JVs in China during the first quarter of 2022.

Notable instances of private–private JVs include the following:

(i) Sony Ericsson, which was established as a JV between Sony Corporation (Japan) and Ericsson (Sweden), merging their expertise in electronics and telecommunications;
(ii) Dow Corning, which was formed through a partnership between Dow Chemical Company (USA) and Corning Incorporated (USA), focusing on silicone product development;
(iii) BMW Brilliance, a collaboration between BMW (Germany) and Brilliance Auto (China), aimed at manufacturing and selling BMW vehicles within China;
(iv) Renault Nissan, the strategic alliance between Nissan and Renault, which encompasses JVs across various markets, enabling the sharing of technology and production resources;
(v) American Motors Corporation's (AMC) collaboration with China to manufacture Jeep-model vehicles in Beijing;
(vi) IKEA virtual store's establishment on the Alibaba e-commerce platform Tmall through a partnership with Alibaba, which enhances IKEA's accessibility for consumers in China and allows for experimentation with new customer engagement strategies.

Prominent instances of public–private partnerships include the following:

(i) PetroCanada, a Canadian hydrocarbon enterprise with a capital structure of 60% public ownership and 40% private ownership;
(ii) Port Darwin's management by the Chinese-owned Landbridge Group, which holds a 99-year lease granted by the Australian government;

(iii) The Mixed Companies of the Orinoco Oil Belt, which is a Venezuelan consortium formed between the State and several transnational corporations for hydrocarbon extraction;
(iv) Red Eléctrica de España, a significant Spanish energy provider that maintains 20% of its shares in public hands, while the remainder is privately owned.

Instances of JVs between governments include partnerships among various governmental bodies, typically aimed at fulfilling particular goals, such as infrastructure enhancement, economic projects, or research and development. Following are some examples:

(i) The Australia New Zealand Leadership Meeting (ANZLM) focusing on collaboration in areas, such as defense and security, trade and investment, development cooperation, sustainability, and cultural exchange;
(ii) Joint efforts in the development of ports, railways, highways, electricity, aviation, and communication infrastructure along the Belt and Road initiative;
(iii) The Global Climate Accords representing UN-led initiatives designed to mitigate the impacts of severe climate change.

Dynamic cooperation between venture parties is a complicated optimization process despite its potential for enhanced joint profits. According to Harvard Business Review and McKinsey & Company, the likelihood of success for JVs is estimated to be only about 50–60%. In addition to fulfilling the requirements of individual rationality and group optimality, a viable collaborative solution must also satisfy the criterion of time consistency. A solution is deemed time-consistent if extending the solution policy to a later initiation and a state resulting from previous optimal actions remains effective (refer to Yeung and Petrosyan, 2004, 2006, 2010, 2016).

Another significant issue concerning JVs is the distribution of cooperative gains. The arrangements for gain-sharing influence the willingness of participants to collaborate and the sustainability of the initiatives. Numerous JV dissolutions have been attributed to dissatisfaction with the gain-sharing agreements. These arrangements are vital components in establishing enduring cooperation. An effective principle for gain-sharing of cooperative gains is essential for securing the consensus of all parties involved in the collaboration and ensuring the sustainability of the JV over time. The disparities among venture participants regarding size, functionality, capability, resources, facilities, and growth potential create substantial challenges in devising gain-sharing arrangements that are satisfactory to all stakeholders. Yeung and Petrosyan (2026) provided an analysis of time-consistent and contributions-guided fair gain-sharing cooperative solutions for JVs.

A well-designed solution addressing the obstacles faced by JVs can significantly enhance the benefits derived from the international division of labor on a global scale and foster an effective international business framework. In the following sections, we construct an innovative cooperative game solution for JV operations, which includes (i) individual and collective rationalities, (ii) a just gain-sharing framework that reflects the contributions of venture partners, and (iii) the consistency of the mutually agreed terms that remain optimal throughout all phases of the venture's lifespan.

2. General Game Framework of Joint Venture

In this section, we first present the benchmark game scenario of the participants before a JV is formed. The corresponding non-cooperative equilibrium is derived as a reference point. Then, we provide the payoff structures of agents after a JV is formed and derive the maximized venture payoff.

2.1. Basic Game Formulation and Benchmark Non-cooperative Equilibria

Consider a JV of n parties/agents with a T-stage planning horizon. The payoff agent i over stage 1 to stage T before forming a JV is

$$\sum_{k=1}^{T} g_k^i(x_k, u_k^1, u_k^2, \ldots, u_k^n)\delta_1^k + q_{T+1}^i(x_{T+1})\delta_1^{T+1}, \tag{1}$$

where $u_k^i \in U_k^i \subset R^{n_i}$ is the vector of strategies of agent $i \in \{1, 2, \ldots, n\} \equiv N$, the vector of state variables of the game at stage k is $x_k \in X \subset R^m$, $q_{T+1}^i(x_{T+1})$ is the terminal payoff of agent i at stage $T + 1$, and δ_1^k is the discount factor from stage 1 to stage k.

The evolution dynamics of the state variables are governed by the vector difference equation:

$$x_{k+1} = f_k(x_k, u_k^1, u_k^2, \ldots, u_k^n), \quad x_1 = x_1^0. \tag{2}$$

When all the agents are acting non-cooperatively, the scenario becomes a dynamic game:

$$\max_{\substack{u_k^i, \\ k \in \{1,2,\ldots,T\}}} \left\{ \sum_{k=1}^{T} g_k^i(x_k, u_k^1, u_k^2, \ldots, u_k^n)\delta_1^k + q_{T+1}^i(x_{T+1})\delta_1^{T+1} \right\} \tag{3}$$

for $i \in N$ subject to dynamics (2).

A frequently used way to characterize and derive a feedback Nash equilibrium of the game is provided in the following theorem.

Theorem 2.1. *A set of strategies $\{u_k^{i**} = \varphi_k^i(x)$ for $k \in \{1, 2, \ldots, T\}$ and $i \in N\}$ provides a feedback Nash equilibrium solution to the game (2)–(3) if the following Hamilton–Jacobi–Bellman (HJB) equations are satisfied:*

$$V^i(T+1, x) = q_{T+1}^i(x)\delta_1^T, \tag{4}$$

$$\begin{aligned} V^i(k, x) = \max_{u_k^i} \{ & g_k^i[x, u_k^{1**}, u_k^{2**}, \ldots, u_k^{i-1**}, u_k^i, u_k^{i+1**}, \ldots, u_k^{n**}]\delta_1^k \\ & + V^i[k+1, f_k[x, u_k^{1**}, u_k^{2**}, \ldots, u_k^{i-1**}, u_k^i, u_k^{i+1**}, \ldots, u_k^{n**}]] \} \end{aligned} \tag{5}$$

for $i \in N$ and $k \in \{1, 2, \ldots, T\}$.

Proof. Invoking the discrete-time dynamic programming technique, $V^i(k, x)$ is the maximized payoff of agent i given the equilibrium strategies of the other $n-1$ players $\{u_k^{i**}(x)$ for $j \in N$ and $j \neq i\}$. Hence, a Nash equilibrium appears. □

2.2. Venture Payoff Structure and Maximized Payoff

Consider the case where these n agents form a JV. The agents in the JV establish connections among themselves and carry out new productive activities that did not exist before. The firms' payoffs would be augmented to include additional gains from activities linked to other agents in the venture. In particular, agent i's payoff becomes

$$\sum_{k=1}^{T}\left(g_k^i(x_k, u_k^1, u_k^2, \ldots, u_k^n) + \sum_{\substack{j\in N\\ j\neq i}} g_k^{i(j)}(x_k, u_k^i, u_k^j)\right)\delta_1^k + q_{T+1}^i(x_{T+1})\delta_1^{T+1} \tag{6}$$

for $i, j \in N$ and $j \neq i$, where $g_k^{i(j)}(x_k, u_k^i, u_k^j) \geq 0$ is the additional gain that agent i receives when connecting with agent j under the JV.

The JV will be dissolved if any participating agent opts out of the venture. As a joint entity, the agents would cooperate and maximize their joint payoff by solving the problem:

$$\max_{\substack{u_k^j, j\in N\backslash i\\ k\in\{1,2,\ldots,T\}}}\left\{\sum_{j=1}^{n}\left[\sum_{k=1}^{T}\left(g_k^j(x_k, u_k^1, u_k^2, \ldots, u_k^n) + \sum_{\substack{\varepsilon\in N\\ \varepsilon\neq j}} g_k^{j(\varepsilon)}(x_k, u_k^j, u_k^\varepsilon)\right)\delta_1^k + q_{T+1}^j(x_{T+1})\delta_1^{T+1}\right]\right\}, \tag{7}$$

subject to dynamics (1).

Invoking Bellman's technique of dynamic programming, the optimal solution to the dynamic optimization problem (1) and (7) can be characterized in the following theorem.

Theorem 2.2. *A set of strategies* $\{u_k^{i*} = \psi_k^i(x)$ *for* $k \in \{1, 2, \ldots, T\}$ *and* $i \in N\}$ *provides an optimal solution to the optimization problem* (1) *and* (7) *if the following Bellman equations are satisfied:*

$$W^N(T+1, x) = \sum_{j=1}^{n} q_{T+1}^j(x)\delta_1^T, \tag{8}$$

$$W^N(k,x) = \max_{u_k^1,u_k^2,\ldots,u_k^n} \left\{ \sum_{j=1}^{n} \left(g_k^j(x_k, u_k^1, u_k^2, \ldots, u_k^n) + \sum_{\substack{\varepsilon \in N \\ \varepsilon \neq j}} g_k^{j(\varepsilon)}(x_k, u_k^j, u_k^\varepsilon) \right) \delta_1^k + W^N[k+1, f_k(x_k, u_k^1, u_k^2, \ldots, u_k^n)] \right\} \tag{9}$$

for $k \in \{1, 2, \ldots, T\}$.

Substituting the optimal control $\{u_k^{i*}$ for $k \in \{1, 2, \ldots, T\}$ and $i \in N\}$ from Theorem 2.2 into the state dynamics (1), one can obtain the dynamics of the cooperative trajectory as

$$x_{k+1} = f_k(x_k, u_k^{1*}, u_k^{2*}, \ldots, u_k^{n*}). \tag{10}$$

We use $\{x_k^*\}_{k=1}^T$ to denote the solution to (10), which is the optimal cooperative path.

3. Fair Gain-sharing in Joint Venture

In a JV, variations in company size can greatly influence control, resource contribution, risk exposure, and strategic impact. Larger companies frequently take the lead in decision-making and profit-sharing matters, whereas smaller firms may benefit from access to resources and markets that would otherwise be unattainable. Generally, larger companies possess greater leverage in negotiations and governance frameworks, while smaller companies have diminished control in exchange for access to capital, technology, or market opportunities. Larger firms typically contribute more capital, infrastructure, and personnel, while smaller partners often provide specialized expertise and insights into local markets. Larger firms are capable of absorbing more financial risks and tend to adopt more aggressive growth strategies, whereas smaller firms are at a higher risk if the venture does not succeed, as their investment constitutes a larger portion of their overall resources. Given the differences among the parties involved, some may receive significantly less gain in comparison to their contributions relative to others in the JV. Profit-sharing presents a complex challenge for partners in a JV. An appropriate gain-sharing principle is essential for ensuring the agreement of all parties to collaborate and for maintaining the sustainability of the JV. A fair gain-sharing principle for cooperative gains is vital for securing the consensus of all parties to collaborate and for the sustainability of the JV. To achieve fair gain-sharing, the distribution of cooperative gains must reflect the contributions made by the parties. To establish a dynamic and fair gain-sharing cooperative solution, we begin with the calculation of the additional value that a specific party contributes to the JV payoff. Subsequently, we derive a fair gain-sharing weight to arrive at a fair gain-sharing solution.

3.1. Marginal Contributions of Agents

The addition to the maximized joint payoff by including agent i is the difference between the maximized JV payoff of the collaborating scheme and the joint payoff without agent i. For notational convenience, we use $N\backslash i$ to denote the set of agents without agent i. Hence, we have to solve the following game and obtain the payoff of the set of agents in $N\backslash i$:

$$\max_{\substack{u_k^j, j\in N\backslash i\\ k\in\{1,2,\ldots,T\}}} \left\{ \sum_{j\in N\backslash i} \left[\sum_{k=1}^{T} \left(g_k^j(x_k, u_k^1, u_k^2, \ldots, u_k^n) + \sum_{\substack{\varepsilon\in N\backslash i\\ \varepsilon\neq j}} g_k^{j(\varepsilon)}(x_k, u_k^j, u_k^\varepsilon) \right) \delta_1^k + q_{T+1}^j(x_{T+1})\delta_1^{T+1} \right] \right\}, \tag{11}$$

$$\max_{u_k^i, k\in\{1,2,\ldots,T\}} \left\{ \sum_{k=1}^{T} g_k^i(x_k, u_k^1, u_k^2, \ldots, u_k^n)\delta_1^k + q_{T+1}^i(x_{T+1})\delta_1^{T+1} \right\} \tag{12}$$

for $i \in N$ subject to

$$x_{k+1} = f_k(x_k, u_k^1, u_k^2, \ldots, u_k^n), \quad x_1 = x_1^0. \tag{13}$$

The game equilibrium payoff of the set of agents in $N \backslash i$, denoted by $W^{(N \backslash i)}(k, x)$, can be characterized by the following theorem.

Theorem 3.1. *A set of strategies* $\{\hat{u}_k^i, u_k^{(N \backslash i)j}$ *for* $j \in N \backslash i\}$ *for* $k \in \{1, 2, \ldots, T\}$, *constitutes a set of feedback Nash equilibria of the game* (11)–(13) *if the following HJB equations are satisfied:*

$$W^{N \backslash i}(T+1, x) = \sum_{j \in N \backslash i} q_{T+1}^j(x_{T+1}) \delta_1^{T+1}, \tag{14}$$

$$\begin{aligned} W^{N \backslash i}(k, x) = \max_{u_k^j, j \in N \backslash i} \Bigg\{ \sum_{j \in N \backslash i} \Bigg(& g_k^j(x, u_k^1, u_k^2, \ldots, u_k^{i-1}, \hat{u}_k^i, u_k^{i+1}, \ldots, u_k^n) \\ & + \sum_{\substack{\varepsilon \in N \backslash i \\ \varepsilon \neq j}} g_k^{j(\varepsilon)}(x_k, u_k^j, u_k^\varepsilon) \Bigg) \delta_1^k \\ & + W^{N \backslash i}[k+1, f_k(x, u_k^1, u_k^2, \ldots, u_k^{i-1}, \hat{u}_k^i, u_k^{i+1}, \ldots, u_k^n)] \Bigg\} \end{aligned} \tag{15}$$

for $k \in \{1, 2, \ldots, T\}$,

$$\hat{V}^i(T+1, x) = q_{T+1}^i(x) \delta_1^{T+1}, \tag{16}$$

$$\begin{aligned} \hat{V}^i(k, x) = \max_{u_k^i} \Big\{ & g_k^i(x, u_k^{(N \backslash i)1}, u_k^{(N \backslash i)2}, \ldots, u_k^{(N \backslash i)i-1}, \\ & u_k^i, u_k^{(N \backslash i)i+1}, \ldots, u_k^{(N \backslash i)n}) \delta_1^k \\ & + \hat{V}^i[k+1, f_k(x, u_k^{(N \backslash i)1}, u_k^{(N \backslash i)2}, \ldots, u_k^{(N \backslash i)i-1}, \\ & \quad u_k^i, u_k^{(N \backslash i)i+1}, \ldots, u_k^{(N \backslash i)n})] \Big\} \end{aligned} \tag{17}$$

for $k \in \{1, 2, \ldots, T\}$.

Proof. Invoking the discrete-time dynamic programming technique, $W^{(N\backslash i)}(k,x)$ is the maximized payoff of $N\backslash i$ given the equilibrium strategies of agent i and $\hat{V}^i(k,x)$ is the maximized payoff of agent i given the equilibrium strategies of agents in $N\backslash i$. Hence, a Nash equilibrium appears. □

3.2. Fair Gain-sharing Weight

In a JV, the participating agents would share the excess of the maximized cooperative JV payoff over the sum of the non-cooperative payoff, i.e.

$$\left[W^N(k,x) - \sum_{j=1}^{n} V^j(k,x)\right]. \tag{18}$$

A sustainable sharing mechanism has to reflect the agents' contributions in the cooperative scheme. To resolve the problem, we develop a gain-sharing weight that reflects the participating agents' contributions. Note that the addition to the maximized joint payoff by including agent i is the difference between the maximized joint payoff and the maximized joint payoff without agent i, i.e.

$$W^N(k,x) - W^{N\backslash i}(k,x) \tag{19}$$

for $i \in N$. To measure agent i's share of contribution to the maximized joint payoff when all agents' marginal contributions are taken into consideration, we adopt the weight

$$\frac{W^N(k,x) - W^{N\backslash i}(k,x)}{\sum_{\varsigma=1}^{n}\left[W^N(k,x) - W^{N\backslash \varsigma}(k,x)\right]} \tag{20}$$

for agent $i \in N$. The weight (20) reflects the relative size of agent i's contribution to the cooperative gain.

4. A Dynamic Fair-sharing Cooperative Solution

In this section, we develop a dynamic fair gain-sharing solution to the dynamic JV (2)–(3). Then, we present a payoff distribution procedure (PDP) of the cooperative gains that leads to the realization of the solution.

4.1. A Time-consistent Fair Gain-sharing Value

Using (18) and (20), we obtain a gain-sharing solution that allocates the excess of total cooperative payoff over the sum of individual payoff proportionally to the contributions of the agents as

$$\xi^i(k, x_k^*) = \frac{W^N(k,x) - W^{N\backslash i}(k,x)}{\sum_{\varsigma=1}^{n} [W^N(k,x) - W^{N\backslash \varsigma}(k,x)]} \times \left[W^N(k,x) - \sum_{j=1}^{n} V^j(k,x) \right] + V^i(k,x) \qquad (21)$$

for $i \in N$, along the cooperative state trajectory $\{x_k^*\}_{k=1}^{T}$.

The fair-sharing value in (21) has the following properties:

(i) It satisfies group optimality in that

$$\sum_{i=1}^{n} \xi^i(k, x_k^*) = W^N(k, x_k^*).$$

(ii) It provides a fair reflection of the contributions of the companies to the cooperative payoff by sharing the excess of cooperative gain over the sum of the non-cooperative payoffs of all companies according to the proportion of the relative size of the agent's contribution to the cooperative gain net of its non-cooperative payoff, i.e.

$$\frac{W^N(k,x_k^*) - W^{N\backslash i}(k,x_k^*)}{\sum_{\varsigma=1}^{n} [W^N(k,x_k^*) - W^{N\backslash \varsigma}(k,x_k^*)]}.$$

(iii) It fulfils time consistency in that the cooperative gain of agent i

$$\xi^i(k, x_k^*) = \frac{W^N(k,x_k^*) - W^{N\backslash i}(k,x_k^*)}{\sum_{\varsigma=1}^{n} [W^N(k,x_k^*) - W^{N\backslash \varsigma}(k,x_k^*)]} \times \left[W^N(k,x_k^*) - \sum_{j=1}^{n} V^j(k,x_k^*) \right] + V^i(k,x_k^*)$$

for $i \in N$ is maintained at each stage $k \in \{1, 2, \ldots T\}$ along the cooperative state trajectory $\{x_k^*\}_{k=1}^{T}$.

(iv) It satisfies individual rationality in that

$$\xi^i(k, x_k^*) \geq V^i(k, x_k^*)$$

for $k \in \{1, 2, \ldots, T\}$ and $i \in N$.

4.2. Time-consistent Payoff Distribution

A PDP leading to the realization of the agreed-upon payoffs $\xi^i(k, x_k^*)$ in (21) in all periods $k \in \{1, 2, \ldots, T\}$ has to be designed. Invoking Yeung and Petrosyan (2010, 2016), the payment scheme $\beta_k^i(x_k^*)$ provides such a PDP that

$$\begin{aligned}\xi^i(k, x_k^*) &= \frac{W^N(k, x_k^*) - W^{N\backslash i}(k, x_k^*)}{\sum_{\varsigma=1}^{n} [W^N(k, x_k^*) - W^{N\backslash \varsigma}(k, x_k^*)]} \\ &\quad \times \left[W^N(k, x_k^*) - \sum_{j=1}^{n} V^j(k, x_k^*) \right] + V^i(k, x_k^*) \\ &= \beta_k^i(x_k^*) + \left[\sum_{\zeta=k}^{T} \beta_\zeta^i(x_\zeta^*)\delta_1^\zeta + q_{T+1}^i(x_{T+1}^*)\delta_1^{T+1} \right] \end{aligned} \tag{22}$$

for $i \in N$ and $k \in \{1, 2, \ldots, T\}$.

Theorem 4.1. *The agreed-upon imputation $\xi(k, x_k^*)$ for $k \in \{1, 2, \ldots, T\}$ along the cooperative trajectory $\{x_k^*\}_{k=1}^T$, can be realized by a payment equaling*

$$\beta_k^i(x_k^*) = (\delta_1^k)^{-1} \left[\xi^i(k, x_k^*) - \xi^i(k+1, f_k(x_k^*, u_k^{1*}, u_k^{2*}, \ldots, u_k^{n*})) \right], \tag{23}$$

given to agent $i \in N$ in stage $k \in \{1, 2, \ldots, T\}$, which will lead to the realization of the Fair-Sharing Dynamic Cooperative Value in formula (27).

Proof. Using (22), one can obtain

$$\xi^i(k+1, x_{k+1}^*) = B_{k+1}^i(x_{k+1}^*)\delta_1^{k+1} + \left\{ \sum_{\zeta=k+2}^{T} B_\zeta^i(x_\zeta^*)\delta_1^\zeta + q_{T+1}^i(x_{T+1}^*)\delta_1^{T+1} \right\}. \quad (24)$$

Substituting (24) into (22) yields

$$\xi^i(k, x_k^*) = \beta_k^i(x_k^*)\delta_1^k + \xi^i(k+1, x_{k+1}^*),$$

which can be expressed as

$$\xi^i(x_k^*) = \beta_k^i(x_k^*)\delta_1^k + \xi^i\left(k+1, f_k(x_k^*, u_k^{1*}, u_k^{2*}, \ldots, u_k^{n*})\right). \quad (25)$$

$\beta_k^i(x_k^*)$ is the target allocation to agent $i \in N$. However, under cooperation, all agents would use the cooperative strategies and the payoff that agent i will directly receive at stage k along the cooperative trajectory equaling $g_k^i(x_k^*, u_k^{1*}, u_k^{2*}, \ldots, u_k^{n*}))$. Therefore, a compensatory payment

$$\pi_k^i(x_k^*) = \beta_k^i(x_k^*) - g_k^i(u_k^{1*}, u_k^{2*}, \ldots, u_k^{n*}) \quad (26)$$

has to be given to agent $i \in N$ in period for $k \in \{1, 2, \ldots, T\}$ to actually yield the cooperative payoff $\xi^i(k, x_k^*)$. □

5. An Illustrative Example

Consider a JV of n firms with a planning horizon of T stages. The vector of state variables of technology at stage k is $x_k = (x_k^1, x_k^2, \ldots, x_k^n) \in X \subset R^n$. The accumulation dynamics of the productive technologies of the firms are

$$x_{t+1}^i = x_t^i + a_t^i u_t^i - \sigma_t^i x_t^i \quad (27)$$

for $x_1^i = x_1^{i(0)}$, $t \in \{1, 2, \ldots, T\}$ and $i \in N$, where u_t^i is the investment expenditure in the productive technology and σ_1^i is the rate of obsolescence.

The payoff function of firm i before forming a JV is

$$\sum_{\tau=1}^{T}\left[p_\tau^{i(i)}v_\tau^{i(i)}(\varpi_\tau^{i(i)}x_\tau^i)^{1/2}-c_\tau^{i(i)}(v_\tau^{i(i)})^2-c_\tau^i(u_\tau^i)^2\right]\delta_1^\tau$$
$$+(q_{T+1}^{i(i)}x_{T+1}^i+M_{T+1}^i)\delta_1^{T+1} \tag{28}$$

for $i\in N$, where $v_\tau^{i(i)}(\varpi_\tau^{i(i)}x_\tau^i)^{1/2}$ is firm i's production function using x_τ^i and factor $v_\tau^{i(i)}$ as inputs, $p_\tau^{i(i)}$ is the price of firm i's output, $c_\tau^{i(i)}(v_\tau^{i(i)})^2$ is the cost of factor $v_\tau^{i(i)}$, and $c_\tau^i(u_\tau^i)^2$ is the cost of investment u_τ^i.

Invoking Theorem 2.1, the non-cooperative equilibrium of the game of maximizing (28) subject to (27) can be characterized by the following.

Corollary 5.1.

$$V^i(T+1,x)=\left(q_{T+1}^{i(i)}x_{T+1}^i+M_{T+1}^i\right)\delta_1^{T+1},$$

$$V^i(t,x)=\max_{u_t^{i(i)},v_t^{i(i)}}\left\{\left(p_t^{i(i)}v_t^{i(i)}(\varpi_t^{i(i)}x_t^i)^{1/2}-c_t^{i(i)}(v_t^{i(i)})^2-c_t^i(u_t^i)^2\right)\delta_1^t\right.$$
$$+V^i\left(t+1,x_{t+1}^{1**},x_{t+1}^{2**},\ldots,x_{t+1}^{i-1**},x_t^i+a_t^iu_t^i\right.$$
$$\left.\left.-\sigma_t^i\,x_t^i,x_{t+1}^{i+1**},\ldots,x_{t+1}^{n**}\right)\right\} \tag{29}$$

for $t\in\{1,2,\ldots,T\}$, *where* $x_{t+1}^{j**}=x_t^j+a_t^ju_t^{j**}-\sigma_t^jx_t^J$ *for* $j\in N$ *and* $j\neq i$.

Performing the indicated maximization in (29), the non-cooperative payoff of the firms can be obtained as follows.

Proposition 5.1. *The non-cooperative payoff can be obtained as*

$$V^i(t,x)=\left(\sum_{j=1}^{n}A_{t+1}^jx_t^j+C_t^{(i)i}\right)\delta_t^t,\quad t\in\{1,2,\ldots,T\}, \tag{30}$$

where

$$A_{T+1}^{(i)i} = q_{T+1}^{i(i)}, \text{ and } C_{T+1}^{(i)i} = M_{T+1}^{i}, \; A_{T+1}^{(i)j} = 0 \text{ for } j \in N \text{ and } j \neq i,$$

$$A_{T+1}^{(i)i} = \frac{(p_t^{i(i)})^2 \varpi_t^{i(i)}}{4c_t^{i(i)}} + \delta_t^{t+1} A_{t+1}^{(i)i}(1-\sigma_t^i),$$

$$A_t^{(i)j} = 0 \text{ for } j \in N \text{ and } j \neq i,$$

$$C_t^{(i)i} = \frac{(A_{t+1}^{(i)i} a_t^i \delta_t^{t+1})^2}{4c_t^i} + C_{t+1}^{i(i)} \text{ for } t \in \{1, 2, \ldots, T\}.$$

Proof. See Appendix A. □

When the firms form a JV, the payoff function of firm i becomes

$$\sum_{\tau=1}^{T} \left[p_\tau^{i(i)} v_\tau^{i(i)} (\varpi_\tau^{i(i)} x_\tau^i)^{1/2} - c_\tau^{i(i)} (v_\tau^{i(i)})^2 - c_\tau^i (u_\tau^i)^2 \right.$$
$$\left. + \sum_{j \in N, j \neq i} \left(p_\tau^{i(j)} v_\tau^{i(j)} (\varpi_\tau^{i(j)} x_\tau^i)^{1/2} - c_\tau^{i(j)} (v_\tau^{i(j)})^2 + \gamma_\tau^{i(j)} x_\tau^j \right) \right] \delta_1^\tau$$
$$+ \sum_{i=1}^{n} \left(q_{T+1}^{i(i)} x_{T+1}^i + M_{T+1}^i \right) \delta_1^{T+1} \tag{31}$$

for $i \in N$, where $p_\tau^{i(j)} v_\tau^{i(j)} (\varpi_\tau^{i(j)} x_\tau^i)^{1/2}$ is firm i's gross revenue from venture activities connected to firm j using x_τ^i and factor $v_\tau^{i(j)}$ as inputs, $c_\tau^{i(j)} (v_\tau^{i(j)})^2$ is the cost of factor $v_\tau^{i(j)}$, and $\gamma_\tau^{i(j)} x_\tau^j$ is the gains from the technology spillover of firm j's technology x_τ^j under the JV.

Invoking Theorem 2.2, the optimal solution to the cooperative game of maximizing (31) subject to (27) can be characterized by the following.

Corollary 5.2.

$$W^N(T+1, x) = \sum_{i=1}^{n} \left(q_{T+1}^{i(i)} x_{T+1}^i + M_{T+1}^i \right) \delta_1^{T+1},$$

$$W^N(t,x) = \max_{\substack{u_t^i, v_t^{i(i)}, v_t^{i(j)} \\ i\in N, j\in N, j\neq i}} \left\{ \sum_{i=1}^{n} \left[p_t^{i(i)} v_t^{i(i)} (\varpi_t^{i(i)} x_t^i)^{1/2} - c_t^{i(i)} (v_t^{i(i)})^2 - c_t^i (u_t^i)^2 \right.\right.$$

$$\left. + \sum_{j\in N, j\neq i} \left(p_t^{i(j)} v_t^{i(j)} (\varpi_t^{i(j)} x_t^i)^{1/2} - c_t^{i(j)} (v_t^{i(j)})^2 + \gamma_t^{i(j)} x_t^j \right) \right] \delta_1^T$$

$$\left. + W^N\left(t+1, x_{t+1}^1, x_{t+1}^2, \ldots, x_{t+1}^n\right) \right\}, \tag{32}$$

where $x_{t+1}^i = x_t^i + a_t^i u_t^i - \sigma_t^i x_t^i$ *for* $i \in N$.

Performing the indicated maximization in (32), the maximized venture profit can be obtained as follows.

Proposition 5.2. *The maximized profit of the joint venture can be obtained as*

$$W^N(T+1,x) = \sum_{i\in N} \left(q_{T+1}^{i(i)} x_{T+1}^i + M_{T+1}^i \right) \delta_1^{T+1}$$

$$= \left(\sum_{i\in N} A_{T+1}^{(N)i} x_{T+1}^i + C_{T+1}^N \right) \delta_1^{T+1},$$

$$W^N(t,x) = \left(\sum_{i\in N} A_t^{(N)i} x_{Tt}^i + C_{Tt}^N \right) \delta_1^t, \quad t \in \{1,2,\ldots,T\}, \tag{33}$$

where

$$A_t^{(N)i} = \frac{(p_t^{i(i)})^2 \varpi_t^{i(i)}}{4c_t^{i(i)}} + \delta_t^{T+1} A_{t+1}^{(N)i} (1-\sigma_t^i)$$

$$+ \delta_t^{t+1} \sum_{j\in N, j\neq i} \left(\frac{(p_t^{i(j)})^2 \varpi_t^{i(j)}}{4c_t^{i(j)}} + \gamma_t^{j(i)} \right) \quad \text{for } i \in N,$$

$$C_t^{(N)} = \sum_{i=1}^{n} \frac{(A_{t+1}^{(N)i} a_t^i \delta_t^{t+1})^2}{4c_t^i} + \delta_t^{t+1} C_{t+1}^N.$$

Proof. See Appendix B. □

The optimal cooperative strategies can be obtained as (see Appendix B)

$$v_t^{\ell(\ell)*} = \left(\frac{p_t^{\ell(\ell)}}{2c_t^{\ell(\ell)}}\right)(\varpi_t^{\ell(\ell)} x_t^{\ell})^{1/2},$$

$$v_t^{\ell(j)*} = \left(\frac{p_t^{\ell(j)}}{2c_t^{\ell(j)}}\right)(\varpi_t^{\ell(j)} x_t^{j})^{1/2} \text{ for } \ell \in N \text{ and } \ell \neq j;$$

$$u_t^{\ell*} = \frac{A_{t+1}^{(N)\ell} a_t^{\ell} \delta_t^{t+1}}{2c_t^{\ell}} \text{ for } \ell \in N \text{ and } t \in \{1, 2, \ldots, T\}. \tag{34}$$

Substituting $u_t^{\ell*}$ into (27), we obtain the optimal accumulation dynamics of the productive technologies of the firms as

$$x_{t+1}^{\ell} = x_t^{\ell} + a_t^{\ell} \frac{A_{t+1}^{(N)\ell} a_t^{\ell} \delta_t^{t+1}}{2c_t^{\ell}} - \sigma_t^{\ell} x_t^{\ell} \tag{35}$$

for $x_1^{\ell} = x_1^{\ell(0)}$ for $\ell \in N$.

We use $\{x_t^*\}_{t=1}^{T+1} = \{x_t^{1*}, x_t^{2*}, \ldots, x_t^{n*}\}_{t=1}^{T+1}$ to denote the solution of (35), which is the optimal venture technology path.

To obtain the marginal contribution of firm i, we have to solve the following dynamic game:

$$\max_{\substack{u_t^{\ell(\ell)}, v_t^{\ell(\ell)}, v_t^{\ell(j)} \\ \ell \in N \backslash i, j \in N \backslash o, j \neq i}} \left\{ \sum_{\tau=1}^{T} \sum_{\ell \in N \backslash i} \left[p_\tau^{\ell(\ell)} v_\tau^{\ell(\ell)} (\varpi_\tau^{\ell(\ell)} x_\tau^{\ell})^{1/2} - c_\tau^{\ell(\ell)} (v_\tau^{\ell(\ell)})^2 - c_\tau^{\ell} (u_\tau^{\ell})^2 \right.\right.$$

$$\left. + \sum_{j \in N \backslash i, j \neq \ell} \left(p_\tau^{\ell(j)} v_\tau^{\ell(j)} (\varpi_\tau^{\ell(j)} x_\tau^{\ell})^{1/2} - c_\tau^{\ell(j)} (v_\tau^{\ell(j)})^2 + \gamma_\tau^{\ell(j)} x_\tau^{j} \right) \right] \delta_1^{\tau}$$

$$\left. + \sum_{\ell \in N \backslash i} \left(q_{T+1}^{\ell(\ell)} x_{T+1}^{\ell} + M_{T+1}^{\ell} \right) \delta_1^{T+1} \right\}, \tag{36}$$

$$\max_{u_t^{i(i)}, v_t^{i(i)}} \left\{ \sum_{\tau=1}^{T} \left[p_\tau^{i(i)} v_\tau^{i(i)} (\varpi_\tau^{i(i)} x_\tau^{i})^{1/2} - c_\tau^{i(i)} (v_\tau^{i(i)})^2 - c_\tau^{i} (u_\tau^{i})^2 \right] \delta_1^{\tau} \right.$$

$$+ \left(q_{T+1}^{i(i)} x_{T+1}^{i} + M_{T+1}^{i} \right) \delta_1^{T+1} \tag{37}$$

subject to (27).

Invoking Theorem 3.1, the non-cooperative equilibrium of the game of (27), (36)–(37) can be characterized by the following.

Corollary 5.3.

$$W^{N\backslash i}(T+1,x)=\sum_{\ell\in N\backslash i}\left(q_{T+1}^{\ell(\ell)}x_{T+1}^{\ell}+M_{T+1}^{\ell}\right)\delta_1^{T+1},$$

$$W^{N\backslash i}(t,x)=\max_{\substack{u_t^{\ell},v_t^{\ell(\ell)},v_t^{\ell(j)}\\ \ell\in N\backslash i,j\in N\backslash i,j\neq\ell}}\left\{\sum_{\ell\in N\backslash i}\left[p_t^{\ell(\ell)}v_t^{\ell(\ell)}(\varpi_t^{\ell(\ell)}x_t^{\ell})^{1/2}\right.\right.$$
$$-c_t^{\ell(\ell)}(v_t^{\ell(\ell)})^2-c_t^{\ell}(u_t^{\ell})^2$$
$$\left.+\sum_{j\in N\backslash i,j\neq\ell}\left(p_\tau^{\ell(j)}v_\tau^{\ell(j)}(\varpi_\tau^{\ell(j)}x_\tau^{\ell})^{1/2}-c_\tau^{\ell(j)}(v_\tau^{\ell(j)})^2+\gamma_\tau^{\ell(j)}x_\tau^{j}\right)\right]\delta_1^{\tau}$$
$$\left.+W^{N\backslash i}\left(t+1,x_{t+1}^1,x_{t+1}^2,\ldots,x_{t+1}^{i-1},\hat{x}_{t+1}^i,x_{t+1}^{i+1},\ldots,x_{t+1}^n\right)\right\}\tag{38}$$

for $t\in\{1,2,\ldots,T\}$, *where*

$$x_{t+1}^{\ell}=x_t^{\ell}+a_t^{\ell}u_t^{\ell}-\sigma_t^{\ell}x_t^{\ell}\text{ for }\ell\in N\backslash i,\ \hat{x}_{t+1}^i=x_t^i+a_t^i\hat{u}_t^i-\sigma_t^ix_t^i;$$

$$\hat{V}^i(T+1,x)=\left(q_{T+1}^{i(i)}x_{T+1}^i+M_{T+1}^i\right)\delta_1^{T+1},$$
$$\hat{V}^i(t,x)=\max_{u_t^{i(i)},v_t^{i(i)}}\left\{\left(p_t^{i(i)}v_t^{i(i)}(\varpi_t^{i(i)}x_t^i)^{1/2}-c_t^{i(i)}(v_t^{i(i)})^2-c_t^i(u_t^i)^2\right)\delta_1^t\right.$$
$$+\hat{V}^i\left(t+1,x_{t+1}^{(N\backslash i)1},x_{t+1}^{(N\backslash i)2},\ldots,x_{t+1}^{(N\backslash i)i-1},\right.$$
$$\left.\left.x_t^i+a_t^iu_t^i-\sigma_t^ix_t^i,x_{t+1}^{(N\backslash i)i+1},\ldots,x_{t+1}^{(N\backslash i)n}\right)\right\}\tag{39}$$

for $t\in\{1,2,\ldots,T\}$, *where* $x_{t+1}^{(N\backslash i)l}=x_t^{\ell}+a_t^{\ell}u_t^{(N\backslash i)\ell}-\sigma_t^{\ell}x_t^{\ell}$ *for* $\ell\in N\backslash i$.

Performing the maximization operator in (38)–(39), the marginal contributions of firm i can be obtained as follows.

Proposition 5.3. *The marginal contributions of firm i is*

$$W^{N\backslash i}(t,x) = \left(\sum_{\ell \in N} A_t^{(N)\ell} x_{Tt}^{\ell} + C_{Tt}^{(N\backslash i)}\right)\delta_1^t, \tag{40}$$

where

$$C_{T+1}^{(N\backslash i)} = \sum_{\ell \in N\backslash i} M_{T+1}^{\ell}, A_{T+1}^{(N\backslash i)\ell} = q_{T+1}^{\ell(\ell)} \textit{ for } \ell \in N\backslash i \text{ and } A_{T+1}^{(N\backslash i)i} = 0.$$

$$\begin{aligned}
A_t^{(N\backslash i)\ell} &= \frac{(p_t^{\ell(\ell)})^2 \varpi_t^{\ell(\ell)}}{4c_t^{\ell(\ell)}} + \delta_t^{t+1} A_{t+1}^{(N\backslash i)\ell}(1-\sigma_t^{\ell}) \\
&\quad + \delta_t^{t+1} \sum_{j \in N\backslash i, j \neq \ell} \left(\frac{(p_t^{\ell(j)})^2 \varpi_t^{\ell(j)}}{4c_t^{\ell(j)}} + \gamma_t^{j(\ell)}\right) \textit{ for } \ell \in N\backslash i, \\
A_t^{(N\backslash i)i} &= 0, \\
C_t^{(N\backslash i)} &= \sum_{\ell \in N\backslash i} \frac{(A_{t+1}^{(N\backslash i)\ell} a_t^{\ell} \delta_t^{t+1})^2}{4c_t^{\ell}} + \delta_t^{t+1} C_{t+1}^{(N\backslash i)} \textit{ for } t \in \{1,2,\ldots,T\}.
\end{aligned}$$

Proof. See Appendix C. □

The dynamically stable fair gain-sharing value can be obtained as

$$\begin{aligned}
\xi^i(t,x_t^*) &= \frac{W^N(t,x_t^*) - W^{N\backslash i}(t,x_t^*)}{\sum_{\varsigma=1}^{n}[W^N(t,x_t^*) - W^{N\backslash \varsigma}(t,x_t^*)]} \\
&\quad \times \left[W^N(t,x_t^*) - \sum_{j=1}^{n} V^j(t,x_t^*)\right] + V^i(t,x_t^*) \\
&= \frac{\sum_{\ell \in N} A_t^{(N)\ell} x_t^{\ell *} + C_t^N - \left[\sum_{\ell \in N\backslash i} A_t^{(N\backslash i)\ell} x_t^{\ell *} + C_t^{(N\backslash i)}\right]}{\sum_{\varsigma=1}^{n}\left(\sum_{\ell \in N} A_t^{(N)\ell} x_t^{\ell *} + C_t^N - \left[\sum_{\gamma \in N\backslash \varsigma} A_t^{(N\backslash \varsigma)\gamma} x_t^{\gamma *} + C_t^{(N\backslash \varsigma)}\right]\right)}
\end{aligned}$$

$$\times \left(\sum_{\ell \in N} A_t^{(N)\ell} x_t^{\ell *} + C_t^N - \sum_{j=1}^{n} (A_t^{(j)j} x_t^{j*} + C_t^{(j)j}) \right) \delta_1^k$$
$$+ (A_t^{(i)i} x_t^{i*} + C_t^{(i)i}) \delta_1^t. \tag{41}$$

The term $\xi^i(t, x_t^*)$ gives the amount of profit allotted to firm i at stage t.

6. Concluding Remarks

The lack of an acceptable gain-sharing principle during the collaboration period influences the participants' decisions on whether to collaborate and impacts the sustainability of the schemes. Numerous JV dissolutions have been attributed to dissatisfaction with the initial gain-sharing arrangements as the business environment evolves over time. This chapter introduces a gain-sharing principle that reflects the contributions of the participants in a JV. An innovative game solution for JV operations is presented with (i) individual rationality and group optimality, (ii) a fair gain-sharing scheme that reflects the contributions of venture partners, and (iii) the consistency of the mutually agreed-upon agreement, which remains optimal at all stages. This solution addresses the challenges encountered by JVs and has the potential to significantly enhance the benefits derived from the division of labor. This marks the first time in which the study of JVs is analysed with a dynamic fair gain-sharing principle, further theoretical advancements, and anticipated practical applications.

JVs have been the subject of extensive research in recent years. Arora *et al.* (2024) analyzed the performance, lifespan, and evolution of joint ventures. Jin and Zhou (2021) explored the influence of ambidextrous innovation strategies on the performance of international joint ventures as well as how parent control asymmetry and cultural distance moderate these effects. Li (2023) investigated cultural friction within international joint ventures. Lee *et al.* (2021) examined the relationship between technological innovation strategies and the pressures from home and host institutions on the formation of international joint ventures by emerging market firms.

Garabedian *et al.* (2025) assessed the effects of a payer–provider joint venture on healthcare value. Sun *et al.* (2021) analyzed the political aspects of innovation in foreign–host–state joint ventures. Lastly, Zhong *et al.* (2023) studied the relationship between CEO organizational identification and the performance of international joint ventures.

The subject of gain-sharing has been a topic of ongoing debate within the academic community. Luo (2009) investigated the pursuit of distributive justice in international joint ventures. Mangel and Useem (2000) analyzed the strategic significance of gain-sharing in the labor market. Mohamed *et al.* (2020) introduced a knowledge-based algorithm for gain-sharing aimed at addressing optimization challenges. Di Foggia and Beccarello (2020) assessed the effects of a gain-sharing cost-reflective tariff on waste management expenses under incentive regulation. O'Byrne and Rajgopal (2022) evaluated employee value added through a novel measure of gain-sharing between labor and capital. Carlan *et al.* (2022) explored cost-effectiveness and gain-sharing scenarios related to the acquisition of a blockchain-based application in the supply chain. Keränen *et al.* (2023) examined gain-sharing within the context of performance-based contracting.

Appendices

Appendix A. Proof of Proposition 5.1

From (29), we obtain

$$\begin{aligned} V^i(T+1,x) &= \left(q^{i(i)}_{T+1} x^i_{T+1} + M^i_{T+1}\right)\delta_1^{T+1} \\ &\equiv \left(A^{i(i)}_{T+1} x^i_{T+1} + C^i_{T+1}\right)\delta_1^{T+1}, \end{aligned} \tag{A.1}$$

yielding $A^{(i)i}_{T+1} = q^{i(i)}_{T+1}$, and $C^{(i)i}_{T+1} = M^i_{T+1}$, $A^{(i)j}_{T+1} = 0$ for $j \in N$ and $j \neq i$.

Using the technique of backward induction, we consider first the last period of operation, i.e. stage T. Using (30), we can obtain the game equilibrium strategies in stage T as

$$v_T^{i(i)} = \left(\frac{p_T^{i(i)}}{2c_T^{i(i)}}\right)(\varpi_T^{i(i)} x^i)^{1/2} \text{ and } u_T^i = \frac{A_{T+1}^{(i)i} a_T^i \delta_T^{T+1}}{2c_T^i}. \tag{A.2}$$

Substituting (A.2) into the stage T equation of (30) yields

$$\begin{aligned} V^i(T,x) = {} & \left(\frac{(p_T^{i(i)})^2}{4c_T^{i(i)}}(\varpi_T^{i(i)} x^i) - \frac{(A_{T+1}^{(i)i} a_T^i \delta_T^{T+1})^2}{4c_T^i}\right)\delta_1^T \\ & + \left[A_{T+1}^{i(i)}\left(x_T^i + a_T^i \frac{A_{T+1}^{(i)i} a_T^i \delta_T^{T+1}}{2c_T^i} - \sigma_T^i x_T^i\right)\right. \\ & \left. + \sum_{\substack{j=1 \\ j\neq i}}^{n} A_{T+1}^{i(j)}\left(x_T^j + a_T^j \frac{A_{T+1}^{(j)j} a_T^j \delta_T^{T+1}}{2c_T^j} - \sigma_T^j x_T^j\right) + C_T^{i(i)}\right] \\ & \delta_1^{T+1} \end{aligned} \tag{A.3}$$

for $\in N$. The RHS of equation (A.3) is an equation system with functions linear in x_T. Hence, $V^i(T,x)$ can be expressed as

$$V^i(T,x) = \left(\sum_{j=1}^{n} A_T^{(i)j} x_T^j + C_T^{(i)i}\right)\delta_1^T, \tag{A.4}$$

where

$$\begin{aligned} A_T^{(i)i} &= \frac{(p_T^{i(i)})^2 \varpi_T^{i(i)}}{4c_T^{i(i)}} + \delta_T^{T+1} A_{T+1}^{(i)i}(1-\sigma_T^i), \\ A_T^{(i)j} &= 0 \text{ for } j \in N \text{ and } j \neq i, \\ C_T^{(i)i} &= \frac{(A_{T+1}^{(i)i} a_T^i \delta_T^{T+1})^2}{4c_T^i} + C_{T+1}^{i(i)}. \end{aligned}$$

Repeating the process for stage $t \in \{T-1, T-2, \ldots, 1\}$ yields

$$V^i(t,x) = \left(\sum_{j=1}^{n} A_t^{(i)j} x_t^j + C_t^{(i)i}\right)\delta_1^t, \quad t \in \{1,2,\ldots T\}, \tag{A.5}$$

where

$$A_t^{(i)i} = \frac{(p_t^{i(i)})^2 \varpi_t^{i(i)}}{4c_t^{i(i)}} + \delta_t^{t+1} A_{t+1}^{(i)i}(1-\sigma_t^i),$$

$$A_t^{(i)j} = 0 \text{ for } j \in N \text{ and } j \neq i,$$

$$C_t^{(i)i} = \frac{(A_{t+1}^{(i)i} a_t^i \delta_t^{t+1})^2}{4c_t^i} + C_{t+1}^{i(i)} \text{ for } t \in \{1, 2, \ldots, T\}.$$

Appendix B. Proof of Proposition 5.2

From (32), we obtain

$$\begin{aligned} W^N(T+1, x) &= \sum_{i=1}^{n} \left(q_{T+1}^{i(i)} x_{T+1}^i + M_{T+1}^i \right) \delta_1^{T+1} \\ &\equiv \left(\sum_{i=1}^{n} A_{T+1}^{(N)i} x_{T+1}^i + C_{T+1}^N \right) \delta_1^{T+1}, \end{aligned} \tag{B.1}$$

yielding $C_{T+1}^N = \sum_{i=1}^n M_{T+1}^i$ and $A_{T+1}^{(N)i} = q_{T+1}^{i(i)}$ for $i \in N$.

Using the technique of backward induction, we consider first the last period of operation, i.e. stage T. Using (32), we can obtain the optimal venture strategies in stage T as

$$\begin{aligned} v_T^{\ell(\ell)} &= \left(\frac{p_T^{\ell(\ell)}}{2c_T^{\ell(\ell)}} \right) (\varpi_T^{\ell(\ell)} x_T^\ell)^{1/2}, \\ v_T^{\ell(j)} &= \left(\frac{p_T^{\ell(j)}}{2c_T^{\ell(j)}} \right) (\varpi_T^{\ell(j)} x_T^j)^{1/2} \text{ for } \ell \in N \text{ and } \ell \neq j; \\ u_T^\ell &= \frac{A_{T+1}^{(N)\ell} a_T^\ell \delta_T^{t+1}}{2c_T^\ell} \quad \text{for } \ell \in N. \end{aligned} \tag{B.2}$$

Substituting (B.2) into the stage T equation of (32) yields

$$\begin{aligned} W^N(T, x) = \sum_{i=1}^{n} &\left[\left(\frac{(p_T^{i(i)})^2}{4c_T^{i(i)}} (\varpi_T^{i(i)} x_T^i) - \frac{(A_{T+1}^{(N)i} a_T^i \delta_T^{T+1})^2}{4c_T^i} \right) \right. \\ &\left. + \sum_{j \in N, j \neq i} \left(\frac{(p_T^{i(j)})^2}{4c_T^{i(j)}} (\varpi_T^{i(j)} x_T^i) + \gamma_T^{i(j)} x_T^j \right) \right] \delta_1^T \end{aligned}$$

$$+\left[\sum_{i=1}^{n} A_{T+1}^{(N)i}\left(x_T^i + a_T^i \frac{A_{T+1}^{(N)i} a_T^i \delta_T^{T+1}}{2c_T^i} - \sigma_T^i x_T^i\right) + C_T^N\right] \times \delta_1^{T+1}. \tag{B.3}$$

The RHS of equation (B.3) is a function linear in x_T. Hence, $W^N(T,x)$ can be expressed as

$$W^N(T,x) = \left(\sum_{j=1}^{n} A_T^{(N)i} x_T^i + C_T^N\right)\delta_1^T, \tag{B.4}$$

where

$$A_T^{(N)i} = \frac{(p_T^{i(i)})^2 \varpi_T^{i(i)}}{4c_T^{i(i)}} + \delta_T^{T+1} A_{T+1}^{(N)i}(1-\sigma_T^i)$$
$$+\delta_T^{T+1} \sum_{j\in N, j\neq i} \left(\frac{(p_T^{i(j)})^2 \varpi_T^{i(j)}}{4c_T^{i(j)}} + \gamma_T^{j(i)}\right) \text{ for } i \in N,$$

$$C_T^{(N)} = \sum_{i=1}^{n} \frac{(A_{T+1}^{(N)i} a_T^i \delta_T^{T+1})^2}{4c_T^i} + \delta_T^{T+1} C_{T+1}^N.$$

Repeating the process for stage $t \in \{T-1, T-2, \ldots, 1\}$ yields

$$W^N(t,x) = \left(\sum_{j=1}^{n} A_{t+1}^{(N)i} x_t^i + C_t^N\right)\delta_1^t, \tag{B.5}$$

where

$$A_t^{(N)i} = \frac{(p_t^{i(i)})^2 \varpi_t^{i(i)}}{4c_t^{i(i)}} + \delta_t^{T+1} A_{t+1}^{(N)i}(1-\sigma_t^i)$$
$$+\delta_t^{t+1} \sum_{j\in N, j\neq i} \left(\frac{(p_t^{i(j)})^2 \varpi_t^{i(j)}}{4c_t^{i(j)}} + \gamma_t^{j(i)}\right) \text{ for } i \in N,$$

$$C_t^{(N)} = \sum_{i=1}^{n} \frac{(A_{t+1}^{(N)i} a_t^i \delta_t^{t+1})^2}{4c_t^i} + \delta_t^{t+1} C_{t+1}^N.$$

Appendix C. Proof of Proposition 5.3

From (38), we obtain

$$\begin{aligned} W^{N\backslash i}(T+1,x) &= \sum_{\ell\in N\backslash i}\left(q_{T+1}^{\ell(\ell)}x_{T+1}^{\ell}+M_{T+1}^{\ell}\right)\delta_1^{T+1} \\ &\equiv \left(\sum_{\ell\in N\backslash i}^{n} A_{T+1}^{(N\backslash i)\ell}x_{T+1}^{\ell}+C_{T+1}^{(N\backslash i)}\right)\delta_1^{T+1}, \quad \text{(C.1)} \end{aligned}$$

yielding $C_{T+1}^{(N\backslash i)i}=\sum_{\ell\in N\backslash i}M_{T+1}^{\ell}$, $A_{T+1}^{(N\backslash i)\ell}=q_{T+1}^{\ell(\ell)}$ for $\ell\in N\backslash i$, and $A_{T+1}^{(N\backslash i)i}=0$.

From (39), we obtain

$$\begin{aligned} \hat{V}^{i}(T+1,x) &= \left(q_{T+1}^{i(i)}x_{T+1}^{i}+M_{T+1}^{i}\right)\delta_1^{T+1} \\ &\equiv \left(\hat{A}_{T+1}^{i(i)}x_{T+1}^{i}+\hat{C}_{T+1}^{i}\right)\delta_1^{T+1}, \quad \text{(C.2)} \end{aligned}$$

yielding $\hat{A}_{T+1}^{(i)i}=q_{T+1}^{i(i)}$, and $\hat{C}_{T+1}^{(i)i}=M_{T+1}^{i}$, $\hat{A}_{T+1}^{(i)j}=0$ for $j\in N$ and $j\neq i$.

Using the technique of backward induction, we consider first the last period of operation, i.e. stage T. Using (38), we can obtain the game equilibrium strategies of the venture $N\backslash i$ in stage T as

$$\begin{aligned} v_T^{\ell(\ell)} &= \left(\frac{p_T^{\ell(\ell)}}{2c_T^{\ell(\ell)}}\right)(\varpi_T^{\ell(\ell)}x_T^{\ell})^{1/2}, \\ v_T^{\ell(j)} &= \left(\frac{p_T^{\ell(j)}}{2c_T^{\ell(j)}}\right)(\varpi_T^{\ell(j)}x_T^{j})^{1/2} \text{ for } \ell\in N\backslash i \text{ and } \ell\neq j; \\ u_T^{\ell} &= \frac{A_{T+1}^{(N\backslash i)\ell}a_T^{\ell}\delta_T^{t+1}}{2c_T^{\ell}} \text{ for } \ell\in N\backslash i. \quad \text{(C.3)} \end{aligned}$$

Using (39), we can obtain the game equilibrium strategies of firm i in stage T as

$$\hat{v}_T^{i(i)}=\left(\frac{p_T^{i(i)}}{2c_T^{i(i)}}\right)(\varpi_T^{i(i)}x_T^{i})^{1/2} \text{ and } \hat{u}_T^{i}=\frac{\hat{A}_{T+1}^{i}a_T^{i}\delta_T^{t+1}}{2c_T^{i}}. \quad \text{(C.4)}$$

Substituting (C.3)–(C.4) into the stage T equation of (38)–(39) yields

$$
\begin{aligned}
W^{N\backslash i}(T,x) = \sum_{\ell\in N\backslash i} & \left[\left(\frac{(p_T^{\ell(\ell)})^2}{4c_T^{\ell(\ell)}} (\varpi_T^{\ell(\ell)} x_T^\ell) - \frac{(A_{T+1}^{(N\backslash i)\ell} a_T^\ell \delta_T^{T+1})^2}{4c_T^\ell} \right) \right. \\
& \left. + \sum_{j\in N\backslash i, j\neq \ell} \left(\frac{(p_T^{\ell(j)})^2}{4c_T^{\ell(j)}} (\varpi_T^{\ell(j)} x_T^\ell) + \gamma_T^{\ell(j)} x_T^j \right) \right] \delta_1^T \\
& + \left[\sum_{\ell\in N\backslash i} A_{T+1}^{(N\backslash i)\ell} \left(x_T^\ell + a_T^\ell \frac{A_{T+1}^{(N\backslash i)\ell} a_T^\ell \delta_T^{T+1}}{2c_T^\ell} - \sigma_T^\ell x_T^\ell \right) \right. \\
& \left. + A_{T+1}^{(N\backslash i)i} \left(x_T^i + a_T^i \frac{\hat{A}_{T+1}^i a_T^i \delta_T^{T+1}}{2c_T^i} - \sigma_T^i x_T^i \right) + C_T^N \right] \\
& \times \delta_1^{T+1};
\end{aligned}
$$

$$
\begin{aligned}
\hat{V}^i(T,x) = & \left(\frac{(p_T^{i(i)})^2}{4c_T^{i(i)}} (\varpi_T^{i(i)} x_T^i) - \frac{(\hat{A}_{T+1}^{(i)i} a_T^i \delta_T^{T+1})^2}{4c_T^i} \right) \delta_1^T \\
& + \left[\hat{A}_{T+1}^{i(i)} \left(x_T^i + a_T^i \frac{\hat{A}_{T+1}^{(i)i} a_T^i \delta_T^{T+1}}{2c_T^i} - \sigma_T^i x_T^i \right) \right. \\
& \left. + \sum_{\substack{j=1 \\ j\neq i}}^n \hat{A}_{T+1}^{i(j)} \left(x_T^j + a_T^j \frac{A_{T+1}^{(N\backslash i)j} a_T^j \delta_T^{T+1}}{2c_T^j} - \sigma_T^j x_T^j \right) + \hat{C}_T^{i(i)} \right] \\
& \times \delta_1^{T+1}.
\end{aligned} \tag{C.5}
$$

The RHS of the equation system is a function linear in x_T. Hence, $W^{N\backslash i}(T,x)$ and $\hat{V}^i(T,x)$ can be expressed as

$$
W^{N\backslash i}(T,x) = \sum_{\ell\in N\backslash i} \left(A_T^{(N\backslash i)\ell} x_T^\ell + C_T^{(N\backslash i)} \right) \delta_1^T, \tag{C.6}
$$

where

$$A_T^{(N\backslash i)\ell} = \frac{(p_T^{\ell(\ell)})^2 \varpi_T^{\ell(\ell)}}{4c_T^{\ell(\ell)}} + \delta_T^{T+1} A_{T+1}^{(N\backslash i)\ell}(1-\sigma_T^\ell)$$
$$+ \delta_T^{T+1} \sum_{j\in N\backslash i, j\neq \ell} \left(\frac{(p_T^{\ell(j)})^2 \varpi_T^{\ell(j)}}{4c_T^{\ell(j)}} + \gamma_T^{j(\ell)} \right) \text{ for } \ell \in N\backslash i,$$

$$C_T^{(N\backslash i)} = \sum_{\ell\in N\backslash i} \frac{(A_{T+1}^{(N\backslash i)\ell} a_T^\ell \delta_T^{T+1})^2}{4c_T^\ell} + \delta_T^{T+1} C_{T+1}^{(N\backslash i)};$$

$$\hat{V}^i(T,x) = \left(\sum_{j=1}^n \hat{A}_T^{(i)j} x_T^j + \hat{C}_T^{(i)i} \right) \delta_1^T, \tag{C.7}$$

where

$$\hat{A}_T^{(i)i} = \frac{(p_T^{i(i)})^2 \varpi_T^{i(i)}}{4c_T^{i(i)}} + \delta_T^{T+1} \hat{A}_{T+1}^{(i)i}(1-\sigma_T^i),$$
$$\hat{A}_T^{(i)j} = 0 \text{ for } j \in N \text{ and } j \neq i,$$
$$\hat{C}_T^{(i)i} = \frac{(\hat{A}_{T+1}^{(i)i} a_T^i \delta_T^{T+1})^2}{4c_T^i} + \hat{C}_{T+1}^{i(i)}.$$

Repeating the process for stage $t \in \{T-1, T-2, \ldots, 1\}$ yields

$$W^{N\backslash i}(t,x) = \left(\sum_{\ell\in N} A_T^{(N\backslash i)\ell} x_T^\ell + C_T^{(N\backslash i)} \right) \delta_1^t, \tag{C.8}$$

where

$$A_t^{(N\backslash i)\ell} = \frac{(p_t^{\ell(\ell)})^2 \varpi_t^{\ell(\ell)}}{4c_t^{\ell(\ell)}} + \delta_t^{t+1} A_{t+1}^{(N\backslash i)\ell}(1-\sigma_t^\ell)$$
$$+ \delta_t^{t+1} \sum_{j\in N\backslash i, j\neq \ell} \left(\frac{(p_t^{\ell(j)})^2 \varpi_t^{\ell(j)}}{4c_t^{\ell(j)}} + \gamma_t^{j(\ell)} \right) \quad \text{for } \ell \in N\backslash i,$$

$$A_t^{(N\backslash i)i} = 0,$$

$$C_t^{(N\backslash i)} = \sum_{\ell \in N\backslash i} \frac{(A_{t+1}^{(N\backslash i)\ell} a_t^{\ell} \delta_t^{t+1})^2}{4c_t^{\ell}} + \delta_t^{t+1} C_{t+1}^{(N\backslash i)};$$

$$\hat{V}^i(t,x) = \left(\sum_{j=1}^{n} \hat{A}_T^{(i)j} x_T^j + \hat{C}_T^{(i)i} \right) \delta_1^t, \tag{C.9}$$

where

$$\hat{A}_T^{(i)i} = \frac{(p_t^{i(i)})^2 \varpi_t^{i(i)}}{4c_t^{i(i)}} + \delta_t^{t+1} \hat{A}_{t+1}^{(i)i} (1 - \sigma_t^i),$$

$$\hat{A}_t^{(i)j} = 0 \text{ for } j \in N \text{ and } j \neq i,$$

$$\hat{C}_t^{(i)i} = \frac{(\hat{A}_{t+1}^{(i)i} a_t^i \delta_t^{t+1})^2}{4c_t^i} + \hat{C}_{t+1}^{i(i)}.$$

References

Arora, A., Inkpen, A., Amawate, V., & Mishra, D. (2024). Performance, longevity and evolution of joint venture: Investigation in the lifecycle of the venture through the case of an international joint venture in India. *South Asian Journal of Business and Management Cases*, 13(3), 297–320. https://doi.org/10.1177/22779779241295264.

Carlan, V., Sys, C. & Vanelslander, T. (2022). Cost-effectiveness and gain-sharing scenarios for purchasing a blockchain-based application in the maritime supply chain. *European Transport Research Review,* 14, 21. https://doi.org/10.1186/s12544-022-00545-2.

Di Foggia, G., & Beccarello, M. (2020). The impact of a gain-sharing cost-reflective tariff on waste management cost under incentive regulation: The Italian case. *Journal of Environmental Management*, 265, 110526. https://doi.org/10.1016/j.jenvman.2020.110526.

Garabedian, L. F., Wharam, J. F., Newhouse, J. P., Lakoma, M., Argetsinger, S., Zhang, F., & Galbraith, A. A. (2025). The impact of a payer–provider joint venture on healthcare value. *Health Services Research*, 60(3): e14400. https://doi.org/10.1111/1475-6773.14400.

Jin, J. L., & Zhou, K. Z. (2021). Is ambidextrous innovation strategy beneficial to international joint venture performance? Evidence from China.

Journal of International Marketing (East Lansing, Mich.), 29(4), 1–21. https://doi.org/10.1177/1069031X211006075.

Keränen, J., Kienzler, M., Salonen, A., Terho, H., & Totzek, D. (2023). Gain-sharing in performance-based contracting: How risk and fairness drive business customers' willingness-to-switch to a gain-sharing arrangement. *Industrial Marketing Management*, 115, 172–184. https://doi.org/10.1016/j.indmarman.2023.09.013.

Lee, J. Y., Xiao, S., & Choi, B. (2021). Unpacking the drivers of emerging market firms' international joint venture formation: The interplay between technological innovation strategies and home- and host-institutional pressures. *Journal of Business Research*, 134, 378–392. https://doi.org/10.1016/j.jbusres.2021.05.049.

Li, C. (2023). Cultural friction in international joint ventures: Whose cultural distance affects market reactions? *Journal of Management*, 49(7), 2288–2317. https://doi.org/10.1177/01492063221104030.

Luo, Y. (2009). From gain-sharing to gain-generation: The quest for distributive justice in international joint ventures. *Journal of International Management*, 15(4), 343–356. https://doi.org/10.1016/j.intman.2008.12.006.

Mangel, R., & Useem, M. (2000). The strategic role of gainsharing. *Journal of Labor Research*, 21, 327–343. https://doi.org/10.1007/s12122-000-1051-y.

Mohamed, A. W., Hadi, A. A., & Mohamed, A. K. (2020). Gaining-sharing knowledge based algorithm for solving optimization problems: A novel nature-inspired algorithm. *International Journal of Machine Learning and Cybernetics*, 11(7), 1501–1529. https://doi.org/10.1007/s13042-019-01053-x.

O'Byrne, S. F., & Rajgopal, S. (2022). Employee value added: A new measure of gain-sharing between labor and capital. *Journal of Applied Corporate Finance*, 34(2), 30–44. https://doi.org/10.1111/jacf.12503.

Sun, P., Deng, Z., & Wright, M. (2021). Partnering with Leviathan: The politics of innovation in foreign–host–state joint ventures. *Journal of International Business Studies*, 52(4), 595–620. https://doi.org/10.1057/s41267-020-00340-y.

Yeung, D. W. K. & Petrosyan, L. A. (2004). Subgame consistent cooperative solution in stochastic differential games. *Journal of Optimization Theory and Applications* 120(3), 651–666.

Yeung, D. W. K. & Petrosyan, L. A. (2006). *Cooperative Stochastic Differential Games,* New York: Springer Verlag, 242pp.

Yeung, D. W. K. & Petrosyan, L. A. (2010). Subgame consistent solutions for cooperative Stochastic Dynamic Games. *Journal of Optimization Theory and Applications*, 145(3), 579–596.

Yeung, D. W. K. & Petrosyan, L. A. (2016). *Subgame Consistent Cooperation: A Comprehensive Treatise*, New York, NY: Springer.

Yeung, D. W. K. & Petrosyan, L. A. (2026). Time-consistent contributions-guided fair gain-sharing cooperative solution for joint ventures. Forthcoming in *Contributions to Game Theory and Management*, Vol. 18.

Zhong, B.-J., Gong, Y., Shenkar, O., Luo, Y., Xiao, Z., & Zhao, S. (2023). Managing the hearts of boundary spanners: CEO organizational identification and international joint venture performance. *Asia Pacific Journal of Management*, 40(1), 87–119. https://doi.org/10.1007/s10490-021-09780-y.

Chapter 5

Fair Gain-Sharing Under Random Exit

"Everything comes to an end. A good bottle of wine, a summer's day, a long-running sitcom, one's life, and eventually our species. The question for many of us is not that everything will come to an end but when. And can we do anything vaguely useful until it does?"
— **Jasper Forde**

"The future is uncertain but the end is always near."
— **Jim Morrison**

In many game situations, the terminal time is not known with certainty. Frequently observed instances of horizontal uncertainties encompass the terms of office for elected officials, contract renewals, the continuation of agreements subject to periodic negotiations, and the lifespan of business enterprises. A notable example is the assumption that business firms typically possess an indefinitely long life expectancy; however, in reality, the average lifespan of contemporary multinational corporations within the Fortune 500 is approximately 50 years. Furthermore, the closure of firms and the transition of political regimes are often not accurately forecasted. The establishment of a fair gain-sharing mechanism is essential for the development of a cooperative framework that takes horizontal uncertainties into account. This chapter examines fair gain-sharing cooperation in the context of random exit. In Section 1, a dynamic game with random duration is formulated, and a dynamic programming approach for addressing inter-temporal issues with a random horizon is developed to provide a basis for resolving the game problem. Section 2 characterizes the non-cooperative equilibrium and the group optimal solution. Section 3 explores fair gain-sharing cooperation under random exit. The relevant subgame-consistent solution and payment mechanism are presented in Section 4. An example of fair gain-sharing under random exit is illustrated in Section 5. Concluding remarks are included in Section 6.

1. Random-horizon Dynamic Games

Dynamic games with a random horizon are studied in game theory when the endpoint of the game is uncertain. Players interact continuously over time, but the game ends at a random stopping time. Random horizon adds realism since many conflicts rarely have fixed durations, for instance:

(i) There are resource allocation games in which companies invest in durable assets, but the market may collapse at an unpredictable time.
(ii) In economics, random horizons manifest in scenarios such as investments made under uncertain market lifetimes or contracts that may terminate unexpectedly.
(iii) Business firms are competing in a market that could potentially shut down at a random time due to regulatory changes or external shocks.
(iv) Trade disputes usually lack a fixed duration for resolution.
(v) Tariff wars may persist, but their ending remains largely unpredictable.
(vi) Military confrontations do not guarantee a definitive conclusion.
(vii) Business operations that necessitate ongoing lease renewals cannot assure the successful renewal of leases in the future.

Dynamic games with random horizons are a crucial tool for analyzing strategic behavior when future endpoints are uncertain. In this section, we begin by defining a class of dynamic games characterized by random durations. Then, we derive a dynamic programming technique to tackle inter-temporal optimization associated with a random horizon, which will serve as the foundation for addressing the ensuing game problem.

1.1. Game Formulation

Consider the n-person dynamic game with $\hat{T}$-stages, where $\hat{T}$ is a random variable with range $\{1, 2, \ldots, T\}$ and corresponding probabilities $\{\theta_1, \theta_2, \ldots, \theta_T\}$. Conditional upon the reaching of stage τ, the probability of the game lasting up to stages $\tau, \tau+1, \ldots, T$ becomes, respectively,

$$\frac{\theta_\tau}{\sum_{\zeta=\tau}^{T} \theta_\zeta}, \frac{\theta_{\tau+1}}{\sum_{\zeta=\tau}^{T} \theta_\zeta}, \ldots, \frac{\theta_T}{\sum_{\zeta=\tau}^{T} \theta_\zeta}.$$

The payoff of player i at stage $k \in \{1, 2, \ldots, T\}$ is $g_k^i[x_k, u_k^1, u_k^2, \ldots, u_k^n]$. When the game ends after stage $\hat{T}$, player i will receive a terminal payment $q_{\hat{T}+1}^i(x_{\hat{T}+1})$ in stage $\hat{T}+1$.

The state space of the game is $X \in R^m$ and the state dynamics of the game is characterized by the difference equation:

$$x_{k+1} = f_k(x_k, u_k^1, u_k^2, \ldots, u_k^n) \tag{1}$$

for $k \in \{1, 2, \ldots, T\} \equiv T$ and $x_1 = x^0$, where $u_k^i \in R^{m_i}$ is the control vector of player i at stage k and $x_k \in X$ is the state.

The objective of player i is

$$\sum_{\hat{T}=1}^{T} \theta_{\hat{T}} \left\{ \sum_{k=1}^{\hat{T}} g_k^i[x_k, u_k^1, u_k^2, \ldots, u_k^n]\delta_1^k + \delta_1^{\hat{T}+1} q_{\hat{T}+1}^i(x_{\hat{T}+1}) \right\} \tag{2}$$

for $i \in \{1, 2, \ldots, n\} \equiv N$.

To solve the game (1)–(2), we first develop a dynamic programming technique for solving a random-horizon problem.

1.2. Dynamic Programming for Random-horizon Optimization

Consider the case when $n = 1$ in the system (1)–(2). The payoff at stage $k \in \{1, 2, \ldots, T\}$ is $g_k[x_k, u_k]$. If the game ends after stage $\hat{T}$, the decision maker will receive a terminal payment $q_{\hat{T}+1}(x_{\hat{T}+1})$ in stage $\hat{T}+1$.

The problem can be formalized as the maximization of the expected payoff:

$$\sum_{\hat{T}=1}^{T} \theta_{\hat{T}} \left\{ \sum_{k=1}^{\hat{T}} g_k[x_k, u_k]\delta_1^k + \delta_1^{\hat{T}+1} q_{\hat{T}+1}(x_{\hat{T}+1}) \right\} \tag{3}$$

subject to the dynamics

$$x_{k+1} = f_k(x_k, u_k), \quad x_1 = \bar{x}_0. \tag{4}$$

Now, consider the case when stage τ has arrived, and the state is x_τ. The problem can be formalized as the maximization of the payoff:

$$\sum_{\hat{T}=\tau}^{T} \frac{\theta_{\hat{T}}}{\sum_{\zeta=\tau}^{T} \theta_{\hat{T}}} \left\{ \sum_{k=\tau}^{\hat{T}} g_k[x_k, u_k]\delta_1^k + \delta_1^{\hat{T}+1} q_{\hat{T}+1}(x_{\hat{T}+1}) \right\}, \tag{5}$$

subject to the dynamics

$$x_{k+1} = f_k(x_k, u_k), \quad x_\tau = x_\tau. \tag{6}$$

We define the value function $V(\tau, x)$ and the set of strategies $\{u_k^* = \psi_k(x)$ for $k \in \{\tau, \tau+1, \ldots, T\}\}$, which provides an optimal solution to (5)–(6) as follows:

$$\begin{aligned} V(\tau, x) = {} & \max_{u_\tau, u_{\tau+1}, \ldots, u_{\hat{T}}} \sum_{\hat{T}=\tau}^{T} \frac{\theta_{\hat{T}}}{\sum_{\zeta=\tau}^{T} \theta_\zeta} \\ & \times \left\{ \sum_{k=\tau}^{\hat{T}} g_k[x_k, u_k]\delta_1^k + \delta_1^{\hat{T}+1} q_{\hat{T}+1}(x_{\hat{T}+1}) \right\} \\ = {} & \sum_{\hat{T}=\tau}^{T} \frac{\theta_{\hat{T}}}{\sum_{\zeta=\tau}^{T} \theta_\zeta} \left\{ \sum_{k=\tau}^{\hat{T}} g_k[x_k^*, u_k^*]\delta_1^k + \delta_1^{\hat{T}+1} q_{\hat{T}+1}(x_{\hat{T}+1}^*) \right\} \end{aligned} \tag{7}$$

for $\tau \in T$, where $x_{k+1}^* = f_k[x_k^*, u_k^*]$, $x_1^* = x_0$.

A theorem characterizing an optimal solution to the random-horizon problem (3)–(4) is provided in the following.

Theorem 1.1. *A set of strategies* $\{u_k^* = \psi_k(x)$ *for* $k \in T\}$ *provides an optimal solution to the problem* (3)–(4) *if there exist functions* $V(k, x)$ *for* $k \in T$*, such that the following recursive relations are satisfied:*

$$
\begin{aligned}
V(T+1, x) &= q_{T+1}(x)\delta_1^T, \\
V(T, x) &= \max_{u_T} \{g_T[x, u_T]\delta_1^T + V[T+1, f_T(x, u_T)]\}, \\
V(\tau, x) &= \max_{u_T} \Bigg\{ g_\tau[x, u_\tau]\delta_1^\tau + \frac{\theta_\tau}{\sum_{\zeta=\tau}^{T} \theta_\zeta} q_{\tau+1}[f_\tau(x, u_\tau)]\delta_1^{\tau+1} \\
&\quad + \frac{\sum_{\zeta=\tau+1}^{T} \theta_\zeta}{\sum_{\zeta=\tau}^{T} \theta_\zeta} V[\tau+1, f_\tau(x, u_\tau)] \Bigg\} \\
&\quad \text{for } \tau \in \{1, 2, \dots, T-1\}.
\end{aligned} \tag{8}
$$

Proof. By definition, the value function at stage $T+1$ is

$$V(T+1, x) = q_{T+1}(x)\delta_1^{T+1}.$$

We first consider the case when the last stage T has arrived.

The problem then becomes

$$\max_{u_T} \left\{ g_T[x, u_T]\delta_1^T + q_{T+1}(x_{T+1})\delta_1^{T+1} \right\}$$

subject to

$$x_{T+1} = f_T(x_T, u_T), \quad x_T = \bar{x}_T. \tag{9}$$

Using $V(T+1, x) = q_{T+1}(x)\delta_1^{T+1}$, the problem in (9) can be formulated as a single-stage problem

$$\max_{u_T} \{g_T[x, u_T]\delta_1^T + V[T+1, f_T(x, u_T)]\},$$

with $x_T = x$.

Hence, we have $V(T, x) = \max_{u_T} \{g_T[x, u_T]\delta_1^T + V[T+1, f_T(x, u_T)]\}$.

Now, consider the problem in stage $\tau \in \{1, 2, \ldots, T-1\}$ in which one has to maximize

$$\sum_{\hat{T}=\tau}^{T} \frac{\theta_{\hat{T}}}{\sum_{\zeta=\tau}^{T} \theta_\zeta} \left\{ \sum_{k=\tau}^{\hat{T}} g_k[x_k, u_k]\delta_1^k + q_{\hat{T}+1}(x_{\hat{T}+1})\delta_1^{\hat{T}+1} \right\}$$

$$= g_\tau[x_\tau, u_\tau]\delta_1^\tau + \frac{\theta_\tau}{\sum_{\zeta=\tau}^{T} \theta_\zeta} q_{\tau+1}(x_{\tau+1})\delta_1^{\tau+1}$$

$$+ \frac{\sum_{\hat{T}=\tau+1}^{T} \theta_{\hat{T}}}{\sum_{\zeta=\tau}^{T} \theta_\zeta} \left\{ \sum_{k=\tau+1}^{\hat{T}} g_k[x_k, u_k]\delta_1^k + q_{\hat{T}+1}(x_{\hat{T}+1})\delta_1^{\hat{T}+1} \right\}$$

$$= g_\tau[x_\tau, u_\tau]\delta_1^\tau + \frac{\theta_\tau}{\sum_{\zeta=\tau}^{T} \theta_\zeta} q_{\tau+1}(x_{\tau+1})\delta_1^{\tau+1} + \frac{\sum_{\zeta=\tau+1}^{T} \theta_\zeta}{\sum_{\zeta=\tau}^{T} \theta_\zeta} \frac{\sum_{\hat{T}=\tau+1}^{T} \theta_{\hat{T}}}{\sum_{\zeta=\tau+1}^{T} \theta_\zeta}$$

$$\times \left\{ \sum_{k=\tau+1}^{\hat{T}} g_k[x_k, u_k]\delta_1^k + q_{\hat{T}+1}(x_{\hat{T}+1})\delta_1^{\hat{T}+1} \right\}. \tag{10}$$

Invoking the definition of $V(\tau+1, x)$ in (7), problem (10) can be expressed as a single-stage problem:

$$\max_{u_T} \left\{ g_\tau[x, u_\tau]\delta_1^\tau + \frac{\theta_\tau}{\sum_{\zeta=\tau}^{T} \theta_\zeta} q_{\tau+1}[f_\tau(x, u_\tau)]\delta_1^{\tau+1} \right.$$
$$\left. + \frac{\sum_{\zeta=\tau+1}^{T} \theta_\zeta}{\sum_{\zeta=\tau}^{T} \theta_\zeta} V[\tau+1, f_\tau(x, u_\tau)] \right\}, \tag{11}$$

with $x_\tau = x$.

Hence, we have

$$V(\tau, x) = \max_{u_T} \left\{ g_\tau[x, u_\tau]\delta_1^\tau + \frac{\theta_\tau}{\sum_{\zeta=\tau}^{T} \theta_\zeta} q_{\tau+1}[f_\tau(x, u_\tau)]\delta_1^{\tau+1} \right.$$
$$\left. + \frac{\sum_{\zeta=\tau+1}^{T} \theta_\zeta}{\sum_{\zeta=\tau}^{T} \theta_\zeta} V[\tau+1, f_\tau(x, u_\tau)] \right\}$$
$$\text{for } \tau \in \{1, 2, \ldots, T-2\}. \tag{12}$$

and Theorem 1.1 follows. □

Theorem 1.1 yields a set of Bellman equations for random-horizon problems (3)–(4).

2. Non-cooperative Equilibria and Group Optimality under Random Horizon

In this section, we investigate the non-cooperative outcome of the random-horizon discrete-time game (1)–(2) and the group optimal condition.

2.1. Non-cooperative Outcome

A feedback Nash equilibrium of the game can be characterized by the following theorem.

Theorem 2.1. *A set of strategies* $\{u_k^{i**} = \varphi_k^i(x)$ *for* $k \in T$ *and* $i \in N\}$ *provides a feedback Nash equilibrium solution to the game* (1)–(2) *if there exist functions* $V^i(k, x)$ *for* $k \in T$ *and* $i \in N$, *such that the following recursive relations are satisfied:*

$$V^i(T+1, x) = q_{T+1}^i(x)\delta_1^{T+1},$$

$$V^i(T, x) = \max_{u_T^i}\Big\{g_T^i[x, u_T^{1**}, u_T^{2**}, \ldots, u_T^{i-1**}, u_T^i, u_T^{i+1*}, \ldots, u_T^{n**}]\delta_1^T + q_{T+1}^i[f_T(x, u_T^{1**}, u_T^{2**}, \ldots, u_T^{i-1**}, u_T^i, u_T^{i+1*}, \ldots, u_T^{n**})]\delta_1^{T+1}\Big\},$$

$$V^i(\tau, x) = \max_{u_T^i}\Bigg\{g_\tau^i[x, u_\tau^{1**}, u_\tau^{2**}, \ldots, u_\tau^{i-1**}, u_\tau^i, u_\tau^{i+1*}, \ldots, u_\tau^{n**}]\delta_1^\tau + \frac{\theta_\tau}{\sum_{\zeta=\tau}^T \theta_\zeta} q_{\tau+1}^i[f_\tau(x, u_\tau^{1**}, u_\tau^{2**}, \ldots, u_\tau^{i-1**}, u_\tau^i, u_\tau^{i+1*}, \ldots, u_\tau^{n**})]\delta_1^{\tau+1} + \frac{\sum_{\zeta=\tau+1}^T \theta_\zeta}{\sum_{\zeta=\tau}^T \theta_\zeta} V^i\left(\tau+1, q_{\tau+1}^i \times [f_\tau(x, u_\tau^{1**}, u_\tau^{2**}, \ldots, u_\tau^{i-1**}, u_\tau^i, u_\tau^{i+1*}, \ldots, u_\tau^{n**})]\right)\Bigg\}$$

$$\text{for } \tau \in \{1, 2, \ldots, T-1\}. \tag{13}$$

Proof. The conditions in (13) show that the random-horizon dynamic programming result in Theorem 1.1 holds for each player given other players' equilibrium strategies. Hence, the conditions of a Nash (1951) equilibrium are satisfied, and Theorem 2.1 follows. □

The set of equations in (13) represents the discrete analogue of the Isaacs-Bellman equations under a random horizon.

Substituting the set of feedback Nash equilibrium strategies $\{u_k^{i**} = \varphi_k^i(x)$ for $i \in N\}$ into the players' payoff yields

$$V^i(\tau, x) = \sum_{\hat{T}=\tau}^{T} \frac{\theta_{\hat{T}}}{\sum_{\zeta=\tau}^{T} \theta_\zeta} \left\{ \sum_{k=\tau}^{\hat{T}} g_k^i[x_k, u_k^{1**}, u_k^{2**}, \ldots, u_k^{n**}(x_k)]\delta_1^k \right.$$

$$\left. + q_{\hat{T}+1}^i(x_{\hat{T}+1})\delta_1^{\hat{T}+1} \right\}, \quad i \in N. \tag{14}$$

The $V^i(\tau, x)$ value function gives the expected game equilibrium payoff to player i from stage τ to the end of the game.

2.2. Group Optimality under Random Horizon

Now, let us examine the scenario in which the players consent to collaborate and allocate the payoff among themselves based on an optimality principle. Achieving the maximum expected joint payoff for the players ensures group optimality in a game where payoffs can be transferred. In order to maximize their expected joint payoff, the players have to solve the discrete-time dynamic programming problem of maximizing the expected payoff:

$$\sum_{j=1}^{n} \sum_{\hat{T}=1}^{T} \theta_{\hat{T}} \left\{ \sum_{k=1}^{\hat{T}} g_k^j[x_k, u_k^1, u_k^2, \ldots, u_k^n]\delta_1^k + q_{\hat{T}+1}^j(x_{\hat{T}+1})\delta_1^{\hat{T}+1} \right\} \tag{15}$$

subject to (1).

Invoking the random-horizon dynamic programming method in Theorem 1.1, we can characterize an optimal solution to the problem (15) and (1) by the following corollary.

Corollary 2.1. *A set of strategies* $\{u_k^{i*} = \psi_k^i(x)$ *for* $k \in \{1, 2, \ldots, T\}$ *and* $i \in N\}$ *provides an optimal solution to the problems* (15) *and*

(1) *if there exist functions* $W(k,x)$ *for* $k \in \{1,2,\ldots,T\}$, *such that the following recursive relations are satisfied:*

$$W^N(T+1,x) = \sum_{j=1}^{n} q^j_{T+1}(x)\delta_1^{T+1},$$

$$W^N(T,x) = \max_{u^1_T,u^2_T,\ldots,u^n_T} \left\{ \sum_{j=1}^{n} g^j_T[x,u^1_T,u^2_T,\ldots,u^n_T]\delta_1^T + q_{T+1}[f_T(x,u_T,u^1_T,u^2_T,\ldots,u^n_T)]\delta_1^T \right\},$$

$$W(\tau,x) = \max_{u^1_\tau,u^2_\tau,\ldots,u^n_\tau} \left\{ \sum_{j=1}^{n} g^j_\tau[x,u^1_\tau,u^2_\tau,\ldots,u^n_\tau]\delta_1^\tau + \frac{\theta_\tau}{\sum_{\zeta=\tau}^{T}\theta_\zeta} q^j_{\tau+1}[f_\tau(x,u^1_\tau,u^2_\tau,\ldots,u^n_\tau)]\delta_1^\tau] + \frac{\sum_{\zeta=\tau+1}^{T}\theta_\zeta}{\sum_{\zeta=\tau}^{T}\theta_\zeta} W^N[\tau+1,f_\tau(x,u^1_\tau,u^2_\tau,\ldots,u^n_\tau)]\delta_1^\tau \right\}$$

$$\text{for } \tau \in \{1,2,\ldots,T-1\}. \tag{16}$$

Substituting the optimal control $\{\psi^i_k(x)$ for $k \in \{1,2,\ldots,T\}$ and $i \in N\}$ into the state dynamics (1), one can obtain the dynamics of the cooperative trajectory as

$$x_{k+1} = f_k[x_k,{}_k^{1**},u_k^{2**},\ldots,u_k^{n**}], \quad x_1 = x^0. \tag{17}$$

We use x^*_k to denote the solution generated by (17).

Using the set of optimal strategies $\{u^{i*}_k = \psi^i_k(x^*_k)$ for $k \in T$ and $i \in N\}$, one can obtain the expected cooperative payoff as

$$W^N(\tau,x) = \sum_{j=1}^{n}\sum_{\hat{T}=\tau}^{T} \frac{\theta_{\hat{T}}}{\sum_{\zeta=\tau}^{T}\theta_\zeta} \left\{ \sum_{k=\tau}^{\hat{T}} g^j_k\left[x^*_k,u^{1*}_k,u^{2*}_k,\ldots,u^{n*}_k\right]\delta_1^k + q^j_{\hat{T}+1}(x^*_{\hat{T}+1})\delta_1^{\hat{T}+1} \right\} \tag{18}$$

for $\tau \in \{1,2,\ldots,t\}$.

3. Fair Gain-sharing Cooperation under Random Exit

To derive a dynamic fair gain-sharing cooperative solution under random exit, we follow the analysis in Chapter 2 and first calculate the additional value of a particular player to the joint cooperative payoff.

3.1. Marginal Contributions of Players

The addition to the maximized joint payoff by including player i is the difference between the maximized expected joint payoff of the collaborating scheme and the maximized expected joint payoff without player i. For notational convenience, we use $N\backslash i$ to denote the set of players without player i. Hence, we have to solve the following game and obtain the expected payoff of the set of players in $N\backslash i$:

$$\max_{\substack{u_k^j, j\in N\backslash i\\ k\in\{1,2,\ldots,T\}}} \left\{ \sum_{j\in N\backslash i} \sum_{\hat{T}=1}^{T} \theta_{\hat{T}} \left(\sum_{k=1}^{\hat{T}} g_k^j \left(x_k, u_k^1, u_k^2, \ldots, u_k^n\right) \delta_1^k \right.\right.$$
$$\left.\left. + q_{\hat{T}+1}^j (x_{\hat{T}+1}) \delta_1^{\hat{T}+1} \right) \right\}, \tag{19}$$

$$\max_{\substack{u_k^i\\ k\in\{1,2,\ldots,T\}}} \left\{ \sum_{\hat{T}=1}^{T} \theta_{\hat{T}} \left(\sum_{k=1}^{\hat{T}} g_k^i \left(x_k, u_k^1, u_k^2, \ldots, u_k^n\right) \delta_1^k \right.\right.$$
$$\left.\left. + q_{\hat{T}+1}^i (x_{\hat{T}+1}) \delta_1^{\hat{T}+1} \right) \right\} \tag{20}$$

subject to dynamics (1).

The game equilibrium payoff of the set of players in $N\backslash i$, denoted by $W^{(N\backslash i)}(k, x)$, can be characterized by the following theorem.

Theorem 3.1. *A set of strategies* $\{\hat{u}_k^i = \hat{\varphi}_k^i(x),\ u_k^{(N\backslash i)j} = \varphi_k^{(N\backslash i)j}(x)$ *for* $j \in N\backslash i\}$ *for* $k \in \{1, 2, \ldots, T\}$ *constitutes a set of feedback Nash* *equilibria of the games* (1) *and* (19)–(20) *if the following HJB equations are satisfied:*

$$W^{N\backslash i}(T+1,x) = \sum_{j\in N\backslash i} q^j_{T+1}(x)\delta_1^{T+1}, \tag{21}$$

$$\begin{aligned} W^{N\backslash i}(T,x) = \max_{u^j_T, j\in N\backslash i} \Bigg\{ \sum_{j=1}^{n} g^j_T[x, u^1_T, u^2_T, \ldots, u^{i-1}_T, \hat{u}^i_T, \\ u^{i+1}_T, \ldots, u^n_T]\delta_1^T + q_{T+1}[f_T(x, u^1_T, u^2_T, \ldots, u^{i-1}_T, \\ \hat{u}^i_T, u^{i+1}_T, \ldots, u^n_T)]\delta_1^T \Bigg\}, \end{aligned} \tag{22}$$

$$\begin{aligned} W^{N\backslash i}(\tau,x) = \max_{u^1_\tau, u^2_\tau, \ldots, u^n_\tau} \Bigg\{ \sum_{j=1}^{n} g^j_\tau[x, u^1_\tau, u^2_\tau, \ldots, u^{i-1}_\tau, \hat{u}^i_\tau, u^{i+1}_\tau, \ldots, \\ u^n_\tau]\delta_1^\tau + \frac{\theta_\tau}{\sum_{\zeta=\tau}^{T}\theta_\zeta} q^j_{\tau+1}\left[f_\tau(x, u^1_\tau, u^2_\tau, \ldots, u^{i-1}_\tau, \hat{u}^i_\tau, \right. \\ \left. u^{i+1}_\tau, \ldots, u^n_\tau)]\delta_1^\tau\right] + \frac{\sum_{\zeta=\tau+1}^{T}\theta_\zeta}{\sum_{\zeta=\tau}^{T}\theta_\zeta} W^N\left[\tau+1, \right. \\ \left. f_\tau(x, u^1_\tau, u^2_\tau, \ldots, u^{i-1}_\tau, \hat{u}^i_\tau, u^{i+1}_\tau, \ldots, u^n_\tau)]\delta_1^\tau\right] \Bigg\} \end{aligned} \tag{23}$$

for $\tau \in \{1, 2, \ldots T-1\}$;

$$\hat{V}^i(T+1,x) = q^i_{T+1}(x)\delta_1^{T+1}, \tag{24}$$

$$\begin{aligned} \hat{V}^i(T,x) = \max_{u^i_T} \Big\{ g^i_T \Big[x, u^{(N\backslash i)1}_T, u^{(N\backslash i)2}_T, \ldots, u^{(N\backslash i)i-1}_T, u^i_T, \\ u^{(N\backslash i)i+1}_T, \ldots, u^{(N\backslash i)n}_T\Big] \delta_1^T \end{aligned}$$

$$+ q^i_{T+1}\left[f_T\left(x, u_T^{(N\backslash i)1}, u_T^{(N\backslash i)2}, \ldots, u_T^{(N\backslash i)i-1}, u^i_T,\right.\right.$$
$$\left.\left.u_T^{(N\backslash i)i+1}, \ldots, u_T^{(N\backslash i)n}\right)\right]\delta_1^{T+1}\Big\}, \tag{25}$$

$$\hat{V}^i(\tau, x) = \max_{u^i_\tau}\Bigg\{g^i_\tau\Big[x, u_\tau^{(N\backslash i)1}, u_\tau^{(N\backslash i)2}, \ldots, u_\tau^{(N\backslash i)i-1}, u^i_\tau,$$
$$u_\tau^{(N\backslash i)i+1}, \ldots, u_\tau^{(N\backslash i)n}\Big]\delta_1^\tau + \frac{\theta_\tau}{\sum_{\zeta=\tau}^{T}\theta_\zeta}q^i_{\tau+1}$$
$$\times\Big[f_\tau(x, u_\tau^{(N\backslash i)1}, u_\tau^{(N\backslash i)2}, \ldots, u_\tau^{(N\backslash i)i-1}, u^i_\tau,$$
$$u_\tau^{(N\backslash i)i+1}, \ldots, u_\tau^{(N\backslash i)n})\Big]\delta_1^{\tau+1} + \frac{\sum_{\zeta=\tau+1}^{T}\theta_\zeta}{\sum_{\zeta=\tau}^{T}\theta_\zeta}V^i$$
$$\times\Big[\tau+1, f_k(x, u_\tau^{(N\backslash i)1}, u_\tau^{(N\backslash i)2}, \ldots, u_\tau^{(N\backslash i)i-1}, u^i_\tau,$$
$$u_\tau^{(N\backslash i)i+1}, \ldots, u_\tau^{(N\backslash i)n})\Big]\Bigg\}$$
$$\text{for } \tau \in \{1, 2, \ldots, T-1\}. \tag{26}$$

Proof. Invoking the discrete-time dynamic programming technique, $W^{(N\backslash i)}(k, x)$ is the maximized payoff of $N\backslash i$ given the equilibrium strategies of player i and $\hat{V}^i(k, x)$ is the maximized payoff of player i given the equilibrium strategies of players in $N\backslash i$. Hence, a Nash equilibrium appears. □

3.2. Fair Gain-sharing Solution

In a cooperation scheme, the participating players would share the excess of the maximized cooperative joint payoff over the sum of the non-cooperative payoff, i.e.,

$$\left[W^N(k, x) - \sum_{j=1}^{n} V^j(k, x)\right]. \tag{27}$$

The addition to the maximized joint payoff by including player i is the difference between the maximized joint payoff and the maximized joint payoff without player i, i.e.,

$$W^N(k,x) - W^{N\backslash i}(k,x) \quad \text{for } i \in N. \tag{28}$$

A fair gain-sharing weight reflecting the share of contribution to the maximized joint payoff can be obtained as

$$\frac{W^N(k,x) - W^{N\backslash i}(k,x)}{\sum_{\varsigma=1}^{n} [W^N(k,x) - W^{N\backslash \varsigma}(k,x)]} \quad \text{for player } i \in N. \tag{29}$$

A dynamically stable fair gain-sharing value can be obtained as

$$\begin{aligned}\xi^i(k,x_k^*) &= \frac{W^N(k,x_k^*) - W^{N\backslash i}(k,x_k^*)}{\sum_{\varsigma=1}^{n} [W^N(k,x_k^*) - W^{N\backslash \varsigma}(k,x_k^*)]} \\ &\quad \times \left[W^N(k,x_k^*) - \sum_{j=1}^{n} V^j(k,x_k^*) \right] + V^i(k,x_k^*) \\ &\quad \text{for } i \in N \end{aligned} \tag{30}$$

along the optimal cooperative trajectory $\{x_k^*\}_{k=1}^{T}$.

4. Subgame-consistent Solutions and Payment Mechanism

To ensure the dynamic stability of a cooperation scheme under random exit, the solution must fulfill the criterion of subgame consistency. Specifically, subgame consistency guarantees that as the game advances, players are guided by the same principle of optimality at every stage, thereby eliminating any incentives to deviate from the previously established optimal behavior. Consequently, for subgame consistency to be achieved, the imputation $\xi(\cdot,\cdot)$ must align with the original optimality along the cooperative trajectory $\{x_k^*\}_{k=1}^{T}$. Under the fair gain-sharing principle, the agreed-upon distribution of payoff is (30).

Crucial to the analysis is the formulation of a payment mechanism so that the imputation in (30) can be realized as the game proceeds.

Following the analysis of Yeung and Petrosyan (2010), we formulate a discrete-time random-horizon payoff distribution procedure

(PDP) so that the agreed-upon imputations (30) can be realized. Let $B_k^i(x_k^*)$ denote the payment that player i will receive at stage k under the cooperative agreement.

The payment scheme involving $B_k^i(x_k^*)$ constitutes a PDP in the sense that along the cooperative trajectory $\{x_k^*\}_{k=1}^T$, the imputation to player i over the stages from k to T can be expressed as

$$\xi^i(\tau, x_\tau^*) = \sum_{\hat{T}=\tau}^{T} \frac{\theta_{\hat{T}}}{\sum_{\zeta=\tau}^{T} \theta_\zeta} \left\{ \sum_{k=\tau}^{\hat{T}} B_k^i(x_k^*)\delta_1^k + q_{\hat{T}+1}^i(x_{\hat{T}+1}^*)\delta_1^{\hat{T}+1} \right\}$$

$$\text{for } i \in N \quad \text{and} \quad k \in T. \tag{31}$$

If the game lasts up to stage T, then at stage $T+1$, player i will receive a terminal payment $q_{T+1}^i(x_{T+1}^*)$, and $B_{T+1}^i(x_{T+1}^*) = 0$. Hence, the imputation $\xi^i(T+1, x_{T+1}^*)$ equals $q_{T+1}^i(x_{T+1}^*)$.

A theorem characterizing a formula for $B_k^i(x_k^*)$ for $k \in T$ and $i \in N$, which yields (32), is provided in the following.

Theorem 4.1. *A payment equaling*

$$\begin{aligned} &B_\tau^i(x_\tau^*) \\ &= \frac{1}{\delta_1^\tau}\left(\xi^i(\tau, x_\tau^*) - \frac{\sum_{\zeta=\tau+1}^{T} \theta_\zeta}{\sum_{\zeta=\tau}^{T} \theta_\zeta} \xi^i(\tau+1, f_\tau[x_\tau^*, u_\tau^{1*}, u_\tau^{2*}, \ldots, u_\tau^{n*}]) \right) \\ &- \frac{\theta_\tau}{\sum_{\zeta=\tau}^{T} \theta_\zeta} q_{\tau+1}^i(f_\tau[x_\tau^*, u_\tau^{1*}, u_\tau^{2*}, \ldots, u_\tau^{n*}])\delta_\tau^{\tau+1} \quad \text{for } i \in N, \end{aligned} \tag{32}$$

given to player i at stage $\tau \in T$ would lead to the realization of the imputation $\xi(\tau, x_\tau^*)$ in (31).

Proof. Using (32), we obtain

$$\begin{aligned} \xi^i(\tau, x_\tau^*) = B_\tau^i(x_\tau^*)\delta_1^\tau + \frac{\theta_\tau}{\sum_{\zeta=\tau}^{T} \theta_\zeta} q_{\tau+1}^i(x_{\tau+1}^*)\delta_1^{\tau+1} \\ + \frac{\sum_{\zeta=\tau+1}^{T} \theta_{\hat{T}}}{\sum_{\zeta=\tau}^{T} \theta_\zeta} \left\{ \sum_{k=\tau+1}^{\hat{T}} B_k^i(x_k^*)\delta_1^k + q_{\hat{T}+1}^i(x_{\hat{T}+1}^*)\delta_1^{\hat{T}+1} \right\} \end{aligned}$$

$$= B_\tau^i(x_\tau^*) + \frac{\theta_\tau}{\sum_{\zeta=\tau}^T \theta_\zeta} q_{\tau+1}^i(x_{\tau+1}^*)\delta_i^{\tau+1}$$

$$+ \frac{\sum_{\zeta=\tau+1}^T \theta_\zeta}{\sum_{\zeta=\tau}^T \theta_\zeta} \frac{\sum_{\zeta=\tau+1}^T \theta_{\hat{T}}}{\sum_{\zeta=\tau+1}^T \theta_\zeta}$$

$$\times \left\{ \sum_{k=\tau+1}^{\hat{T}} B_k^i(x_k^*)\delta_1^k + q_{\hat{T}+1}^i(x_{\hat{T}+1}^*)\delta_1^{\hat{T}+1} \right\}. \tag{33}$$

Invoking $x_{\tau+1}^* = f_\tau[x_\tau^*, u_\tau^{1*}, u_\tau^{2*}, \ldots, u_\tau^{n*}]$ and the definition of $\xi^i(\tau, x_\tau^*)$ in (32), we can express (33) as

$$\xi^i(\tau, x_\tau^*) = B_\tau^i(x_\tau^*)\delta_1^\tau + \frac{\theta_\tau}{\sum_{\zeta=\tau}^T \theta_\zeta} q_{\tau+1}^i(f_\tau[x_\tau^*, u_\tau^{1*}, u_\tau^{2*}, \ldots, u_\tau^{n*}])\delta_1^{\tau+1}$$

$$+ \frac{\sum_{\zeta=\tau+1}^T \theta_\zeta}{\sum_{\zeta=\tau}^T \theta_\zeta} \xi^i(\tau+1, f_\tau[x_\tau^*, u_\tau^{1*}, u_\tau^{2*}, \ldots, u_\tau^{n*}]). \tag{34}$$

From (34), one can obtain (32). Hence, Theorem 4.1 follows. □

Note that the PDP $B_\tau^i(x_\tau^*)$ in (33) would give rise to the agreed-upon stable fair gain-sharing value:

$$\xi^i(k, x_k^*) = \frac{W^N(k, x_k^*) - W^{N\backslash i}(k, x_k^*)}{\sum_{\varsigma=1}^n [W^N(k, x_k^*) - W^{N\backslash \varsigma}(k, x_k^*)]}$$

$$\times \left[W^N(k, x_k^*) - \sum_{j=1}^n V^j(k, x_k^*) \right] + V^i(k, x_k^*),$$

for $i \in N$.

Therefore, subgame consistency is satisfied.

When all players are using the cooperative strategies, the payoff that player i will directly receive at stage $k \in T$ is

$$g_k^i[x_\tau^*, u_\tau^{1*}, u_\tau^{2*}, \ldots, u_\tau^{n*}].$$

However, according to the agreed-upon imputation, player i is to receive $B_k^i(x_k^*)$ at stage k. Therefore, a side payment

$$\bar{\omega}_k^i(x_k^*) = B_k^i(x_k^*) - g_k^i[x_\tau^*, u_\tau^{1*}, u_\tau^{2*}, \ldots, u_\tau^{n*}] \quad \text{for } k \in T \quad \text{and} \quad i \in N \tag{35}$$

will be given to player i.

5. An Illustration in Fair Sharing under Random Exit

We adopt the game structure of the game of random duration lease in Yeung and Petrosyan (2016). Consider the case when two firms are given the lease to extract a renewable resource from a resource pool. The lease for resource extraction has to be renewed after each stage (year) for up to a maximum of four stages. At stage 1, it is known that the probabilities that the lease will be $1, 2, 3$ or 4 years long are, respectively, $\theta_1, \theta_2, \theta_3$ and θ_4. Conditional upon the reaching of stage $\tau > 1$, the probability of the game lasting up to stages $\tau, \tau+1$, to 4 is

$$\frac{\theta_\tau}{\sum_{\zeta=\tau}^{4}\theta_\zeta}, \quad \frac{\theta_{\tau+1}}{\sum_{\zeta=\tau}^{4}\theta_\zeta}, \quad \frac{\theta_4}{\sum_{\zeta=\tau}^{4}\theta_\zeta}.$$

Let u_k^i denote the amount of resource extraction of firm i at stage k for $i \in \{1, 2\}$ and $x_k \in X \subset R^+$ the size of the resource stock at stage k. The extraction cost for firm $i \in \{1, 2\}$ depends on the quantity of resource extracted u_k^i, the resource stock size x_k, and cost parameters c_i. In particular, the extraction cost for firm i at stage k is $c_i(u_k^i)^2/x_k$. The price of the resource is P.

The profits that firm 1 and firm 2 will obtain at stage k are, respectively,

$$\left[\boldsymbol{P u_k^1 - \frac{c_1}{x_k}(u_k^1)^2}\right] \quad \text{and} \quad \left[\boldsymbol{P u_k^2 - \frac{c_2}{x_k}(u_k^2)^2}\right]. \tag{36}$$

If the game ends after stage $\boldsymbol{\tau}$, a terminal payment $\boldsymbol{q^i x_{\tau+1}}$ will be received by firm $\boldsymbol{i}$ in stage $\boldsymbol{\tau+1}$. The growth dynamics of the resource is governed by the difference equation:

$$x_{k+1} = x_k + a - bx_k - \sum_{j=1}^{2} u_k^j, \quad x_1 = x^0 \tag{37}$$

for $k \in \{1, 2, 3, 4\}$.

In particular, there is an extraction constraint $u_k^1 + u_k^2 \leq (1-b)x_k + a$. The discount rate is r. The objective of extractor

$i \in \{1,2\}$ is to maximize the present value of the expected stream of future profits:

$$\sum_{\hat{T}=1}^{4} \boldsymbol{\theta}_{\hat{T}} \left\{ \sum_{k=1}^{\hat{T}} \left[\boldsymbol{P u_k^i} - \frac{\boldsymbol{c_i}}{\boldsymbol{x_k}} (\boldsymbol{u_k^i})^2 \right] \left(\frac{\mathbf{1}}{\mathbf{1}+\boldsymbol{r}} \right)^{\boldsymbol{k}-\mathbf{1}} + \boldsymbol{q^i x}_{\hat{T}+1} \left(\frac{\mathbf{1}}{\mathbf{1}+\boldsymbol{r}} \right)^{\hat{T}} \right\}, \tag{38}$$

subject to (37).

Invoking Theorem 2.1, one can characterize the non-cooperative equilibrium strategies in a feedback solution for game (37)–(38). In particular, a set of strategies $\{\boldsymbol{u_k^{i**}} = \boldsymbol{\varphi_k^i(x)}$ for $\boldsymbol{k} \in \{\mathbf{1},\mathbf{2},\mathbf{3},\mathbf{4}\}$ and $\boldsymbol{i} \in \{\mathbf{1},\mathbf{2}\}\}$ provides a Nash equilibrium solution to the game (37)–(38) if there exist functions $\boldsymbol{V^i(k,x)}$ for $\boldsymbol{i} \in \{\mathbf{1},\mathbf{2}\}$ and $\boldsymbol{k} \in \{\mathbf{1},\mathbf{2},\mathbf{3},\mathbf{4}\}$, such that the following recursive relations are satisfied:

$$V^i(5,x) = q^i x \left(\frac{1}{1+r} \right)^4,$$

$$V^i(4,x) = \max_{u_4^i} \left\{ \left[P u_4^i - \frac{c_i}{x} (u_4^i)^2 \right] \left(\frac{1}{1+r} \right)^3 + q^i [x + a - bx - u_4^i - u_4^{j**}] \left(\frac{1}{1+r} \right)^4 \right\},$$

$$V^i(\tau,x) = \max_{u_\tau^i} \left\{ \left[P u_\tau^i - \frac{c_i}{x} (u_\tau^i)^2 \right] \left(\frac{1}{1+r} \right)^{\tau-1} + \frac{\theta_\tau}{\sum_{\zeta=\tau}^{T} \theta_\zeta} q^i [x + a - bx - u_\tau^i - u_\tau^{j**}] \left(\frac{1}{1+r} \right)^{\tau} + \frac{\sum_{\zeta=\tau+1}^{T} \theta_\zeta}{\sum_{\zeta=\tau}^{T} \theta_\zeta} V^i[\tau+1, x + a - bx - u_\tau^i - u_\tau^{j**}] \right\}$$

$$\text{for } \tau \in \{1,2,3\}. \tag{39}$$

Performing the indicated maximization in (39) yields

$$
\begin{aligned}
&\left(P-\frac{2c_i u_4^{i**}}{x}\right)-q^i\left(\frac{1}{1+r}\right)=0,\\
&\left(P-\frac{2c_i u_\tau^{i**}}{x}\right)\left(\frac{1}{1+r}\right)^{\tau-1}-\frac{\theta_\tau}{\sum_{\zeta=\tau}^{T}\theta_\zeta}q^i\left(\frac{1}{1+r}\right)^{\tau}\\
&\quad-\frac{\sum_{\zeta=\tau+1}^{T}\theta_\zeta}{\sum_{\zeta=\tau}^{T}\theta_\zeta}V^i_{x_{\tau+1}}[\tau+1,x+a-bx-u_\tau^{i**}-u_\tau^{j**}]=0\\
&\text{for}\quad \tau\in\{1,2,3\}.
\end{aligned}
\tag{40}
$$

From (40), the game equilibrium strategies can be expressed as

$$
\begin{aligned}
\boldsymbol{u_4^{i**}} &= \left[\boldsymbol{P-q^i\left(\frac{1}{1+r}\right)}\right]\frac{\boldsymbol{x}}{\boldsymbol{2c_i}}\quad \text{for } \boldsymbol{i\in\{1,2\}}.\\
\boldsymbol{u_\tau^{i**}} &= \left[\boldsymbol{P-\frac{\theta_\tau}{\sum_{\zeta=\tau}^{T}\theta_\zeta}q^i\left(\frac{1}{1+r}\right)-\frac{\sum_{\zeta=\tau+1}^{T}\theta_\zeta}{\sum_{\zeta=\tau}^{T}\theta_\zeta}}\right.\\
&\quad\left.\boldsymbol{\times V^i_{x_{\tau+1}}[\tau+1,x+a-bx-u_\tau^{i**}-u_\tau^{j**}](1+r)^{\tau-1}}\right]\frac{\boldsymbol{x}}{\boldsymbol{2c_i}}
\end{aligned}
\tag{41}
$$

for $i\in\{1,2\}$ and $\tau\in\{1,2,3\}$.

The value function $V^i(\tau,x)$ indicating the game equilibrium payoff of firm $i\in\{1,2\}$ and $\tau\in\{1,2,3,4\}$ can be obtained as follows.

Proposition 5.1.

$$
V^i(\tau,x)=[A_\tau^i x+C_\tau^i]\quad \text{for } i\in\{1,2\}\quad\text{and}\quad \tau\in\{1,2,3,4\}, \tag{42}
$$

where $\boldsymbol{A_4^i}$ *for* $\boldsymbol{i\in\{1,2\}}$ *are given in* (A.2) *of Appendix A and* $\boldsymbol{C_4^i}=\boldsymbol{q^i a}$, *and* $\boldsymbol{A_\tau^i}$ *and* $\boldsymbol{C_\tau^i}$ *for* $\boldsymbol{i\in\{1,2\}}$ *and* $\boldsymbol{\tau\in\{1,2,3\}}$ *are given in* (A.4) *of Appendix A.*

Proof. See Appendix A. □

Now, consider the case when the extractors agree to maximize their expected joint profit and share the excess of cooperative gains over their expected non-cooperative payoffs equally. To maximize their expected joint payoff, they solve the problem of maximizing

$$\sum_{j=1}^{2}\sum_{\hat{T}=1}^{4}\theta_{\hat{T}}\left\{\sum_{k=1}^{\hat{T}}\left[Pu_k^j-\frac{c_j}{x_k}(u_k^j)^2\right]\left(\frac{1}{1+r}\right)^{k-1}q^j x_{\hat{T}+1}\left(\frac{1}{1+r}\right)^{\hat{T}}\right\} \tag{43}$$

subject to (37).

Invoking Theorem 1.1, one can characterize the optimal controls in the dynamic programming problems (37) and (43). In particular, a set of control strategies $\{\psi_k^i(x)$ for $k\in\{1,2,3,4\}$ and $i\in\{1,2\}\}$ provides an optimal solution to the control problem (37) and (43) if there exist functions $W(k,x):R\rightarrow R$ for $k\in\{1,2,3,4\}$, such that the following recursive relations are satisfied:

$$W(5,x)=\sum_{j=1}^{n}q_{T+1}^j(x),$$

$$W^N(4,x)=\max_{u_4^1,u_4^2}\left\{\sum_{j=1}^{n}\left[\left[Pu_4^j-\frac{c_j}{x}(u_4^j)^2\right]\left(\frac{1}{1+r}\right)^3\right.\right.$$
$$\left.\left.+\;q^j[x+a-bx-u_4^1-u_4^2]\left(\frac{1}{1+r}\right)^4\right]\right\},$$

$$W^N(\tau,x)=\max_{u_\tau^1,u_\tau^2}\left\{\sum_{j=1}^{n}\left[\left[Pu_\tau^j-\frac{c_j}{x}(u_\tau^j)^2\right]\left(\frac{1}{1+r}\right)^{\tau-1}\right.\right.$$
$$\left.+\;\frac{\theta_\tau}{\sum_{\zeta=\tau}^{T}\theta_\zeta}q^j[x+a-bx-u_\tau^1-u_\tau^2]\left(\frac{1}{1+r}\right)^{\tau}\right]$$
$$\left.+\;\frac{\sum_{\zeta=\tau+1}^{T}\theta_\zeta}{\sum_{\zeta=\tau}^{T}\theta_\zeta}W[\tau+1,x+a-bx-u_\tau^1-u_\tau^1]\right\}$$

for $\tau\in\{1,2,3\}$. (44)

Performing the indicated maximization in (44) yields

$$\left(P-\frac{2c_i\psi_4^i(x)}{x}\right)-(q^1+q^2)\left(\frac{1}{1+r}\right)=0 \quad \text{for } i\in\{1,2\},$$

$$\left(P-\frac{2c_i\psi_4^i(x)}{x}\right)\left(\frac{1}{1+r}\right)^{\tau-1}-\frac{\theta_\tau}{\sum_{\zeta=\tau}^{T}\theta_\zeta}(q^1+q^2)\left(\frac{1}{1+r}\right)^{\tau}$$
$$-\frac{\sum_{\zeta=\tau+1}^{T}\theta_\zeta}{\sum_{\zeta=\tau}^{T}\theta_\zeta}W^N_{x_{\tau+1}}[\tau+1,x+a-bx-\psi_\tau^1(x)-\psi_\tau^2(x)]=0 \tag{45}$$

for $i\in\{1,2\}$ and $\tau\in\{1,2,3\}$.

From (45), the optimal cooperative strategies can be expressed as

$$\boldsymbol{u_4^{i*}(x)=\left[P-(q^1+q^2)\left(\frac{1}{1+r}\right)\right]\frac{x}{2c_i}} \quad \text{for } \boldsymbol{i\in\{1,2\}},$$

$$\boldsymbol{u_\tau^{i*}(x)=\left[P-\frac{\theta_\tau}{\sum_{\zeta=\tau}^{T}\theta_\zeta}(q^1+q^2)\left(\frac{1}{1+r}\right)-\frac{\sum_{\zeta=\tau+1}^{T}\theta_\zeta}{\sum_{\zeta=\tau}^{T}\theta_\zeta}\right.}$$
$$\boldsymbol{\left.\times W_{x_{\tau+1}}[\tau+1,x+a-bx-\psi_\tau^1(x)-\psi_\tau^2(x)](1+r)^{\tau-1}\right]\frac{x}{2c_i}}$$

for $\boldsymbol{i\in\{1,2\}}$ and $\boldsymbol{\tau\in\{1,2,3\}}$. (46)

The joint payoff of the firms under cooperation can be obtained as follows.

Proposition 5.2. *The value function indicating the maximized joint payoff is*

$$W^N(\tau,x)=[A_\tau x+C_\tau] \quad \text{for } \tau\in\{1,2,3,4\}, \tag{47}$$

where $\boldsymbol{A_4}$ *is given in* (B.2) *of Appendix B and* $\boldsymbol{C_4=(q^1+q^2)a}$, *and* $\boldsymbol{A_\tau}$ *and* $\boldsymbol{C_\tau}$ *for* $\boldsymbol{\tau\in\{1,2,3\}}$ *are given in* (B.4) *of Appendix B.*

Proof. See Appendix B. □

Using (46) and Proposition 5.2, the optimal cooperative strategies of the players can be expressed as

$$\boldsymbol{u_4^{i*}(x) = \left[P - (q^1 + q^2)\left(\frac{1}{1+r}\right)\right]\frac{x}{2c_i}} \quad \text{for } \boldsymbol{i \in \{1, 2\}},$$

$$\boldsymbol{u_\tau^{i*}(x) = \left[P - \frac{\theta_\tau}{\sum_{\zeta=\tau}^{T}\theta_\zeta}(q^1 + q^2)\left(\frac{1}{1+r}\right)\right.}$$
$$\boldsymbol{\left. - \frac{\sum_{\zeta=\tau+1}^{T}\theta_\zeta}{\sum_{\zeta=\tau}^{T}\theta_\zeta}A_{x_{\tau+1}}(1+r)^{\tau-1}\right]\frac{x}{2c_i}} \tag{48}$$

for $i \in \{1, 2\}$ and $\tau \in \{1, 2, 3\}$.

Substituting $\boldsymbol{u_k^{i*}(x)}$ from (48) into (37) yields the optimal cooperative state trajectory:

$$x_5 = x_4 + a - bx_4 - \sum_{j=1}^{2}\left[P - (q^1 + q^2)\left(\frac{1}{1+r}\right)\right]\frac{x_4}{2c_j},$$

$$x_{\tau+1} = x_\tau + a - bx_\tau - \sum_{j=1}^{2}\left[P - \frac{\theta_\tau}{\sum_{\zeta=\tau}^{T}\theta_\zeta}(q^1 + q^2)\left(\frac{1}{1+r}\right)\right.$$
$$\left. - \frac{\sum_{\zeta=\tau+1}^{T}\theta_\zeta}{\sum_{\zeta=\tau}^{T}\theta_\zeta}A_{x_{\tau+1}}(1+r)^{\tau-1}\right]\frac{x_\tau}{2c_j}$$
$$\text{for } \tau \in \{1, 2, 3\} \quad \text{and} \quad x_1 = x^0. \tag{49}$$

Dynamics (49) is a linear first-order difference equation that is readily solvable by standard techniques. Let $\{x_k^* \text{ for } k \in \{1, 2, \ldots, 5\}\}$ denote the solution to (49).

To derive a dynamic fair gain-sharing cooperative solution under random exit, we have to obtain the marginal payoffs of players $W^{N\backslash i}(\tau, x)$ by solving the game:

$$\max_{\substack{u_k^j, j \in N\backslash i \\ k \in \{1,2,\ldots,T\}}} \left\{\sum_{j \in N\backslash i}\sum_{\hat{T}=1}^{T}\theta_{\hat{T}}\left(\sum_{k=1}^{\hat{T}}\left[Pu_k^j - \frac{c_j}{x_k}(u_k^j)^2\right]\left(\frac{1}{1+r}\right)^{k-1}\right.\right.$$
$$\left.\left. + q^j x_{\hat{T}+1}\left(\frac{1}{1+r}\right)^{\hat{T}}\right)\right\}, \tag{50}$$

$$\max_{\substack{u_k^i \\ k\in\{1,2,\ldots,T\}}} \left\{\sum_{\hat{T}=1}^{4} \theta_{\hat{T}} \left\{\sum_{k=1}^{\hat{T}} \left[Pu_k^i - \frac{c_i}{x_k}(u_k^i)^2\right] \left(\frac{1}{1+r}\right)^{k-1} \right.\right.$$
$$\left.\left. + q^i x_{\hat{T}+1} \left(\frac{1}{1+r}\right)^{\hat{T}} \right\}\right\} \tag{51}$$

subject to dynamics (37).

Since there are only two extractors, the games (37) and (50)–(51) is the non-cooperative game between these extractors. Therefore, $W^{N\backslash i}(\tau, x) = V^j(\tau, x)$ and $W^{N\backslash j}(\tau, x) = V^i(\tau, x)$.

A dynamically stable fair gain-sharing value can be obtained as

$$\begin{aligned}
\xi^i(k, x_k^*) &= \frac{W^N(k, x_k^*) - W^{N\backslash i}(k, x_k^*)}{\sum_{\varsigma=1}^{2} [W^N(k, x_k^*) - W^{N\backslash \varsigma}(k, x_k^*)]} \\
&\quad \times \left[W^N(k, x_k^*) - \sum_{j=1}^{2} V^j(k, x_k^*)\right] + V^i(k, x_k^*) \\
&= \frac{A_k x_k^* + C_k - (A_k^i x_k^* + C_k^i)}{\sum_{\varsigma=1}^{2} [(A_k x_k^* + C_k) - (A_k^\varsigma x_k^* + C_k^\varsigma)]} \\
&\quad \times \left[A_k x_k^* + C_k - \sum_{\varsigma=1}^{2} (A_k^\varsigma x_k^* + C_k^\varsigma)\right] + (A_k^i x_k^* + C_k^i) \\
&\quad \text{for } i \in \{1, 2\}.
\end{aligned} \tag{52}$$

Invoking Theorem 4.1, a payment equaling

$$\begin{aligned}
&B_\tau^i(x_\tau^*) \\
&= \frac{1}{\delta_1^\tau} \left(\xi^i(\tau, x_\tau^*) - \frac{\sum_{\zeta=\tau+1}^{T} \theta_\zeta}{\sum_{\zeta=\tau}^{T} \theta_\zeta} \xi^i(\tau+1),\ f_\tau[x_\tau^*, u_\tau^{1*}, u_\tau^{2*}, \ldots, u_\tau^{n*}]\right) \\
&\quad - \frac{\theta_\tau}{\sum_{\zeta=\tau}^{T} \theta_\zeta} q_{\tau+1}^i (f_\tau[x_\tau^*, u_\tau^{1*}, u_\tau^{2*}, \ldots, u_\tau^{n*}]) \delta_\tau^{\tau+1} \quad \text{for } i \in N,
\end{aligned} \tag{53}$$

given to player i at stage $\tau \in \{1, 2, 3, 4\}$ would lead to the realization of the imputation $\xi(\tau, x_\tau^*)$ in (52).

6. Concluding Remarks

Petrosyan and Murzov (1966) were the first to formulate the Bellman Isaacs equations for zero-sum differential games under a random horizon. Subsequently, Petrosyan and Shevkoplyas (2003) provided the initial analysis of dynamically consistent solutions for cooperative differentials with a random duration. In 2010, Shevkoplyas examined the Shapley value within cooperative differential games that operate under a random horizon in an infinite horizon context. Yeung and Petrosyan (2011) developed subgame-consistent cooperative solutions for discrete-time dynamic games featuring random horizons and derived new formulations of the Bellman and Isaacs–Bellman equations. Ferreira *et al.* (2019) investigated stochastic differential games in which the planning horizon is treated as a random variable, employing dynamic programming and viscosity solutions to demonstrate the existence and uniqueness of the game's value. Zaremba *et al.* (2020) explored differential games with a random duration that is governed by a discontinuous cumulative distribution function. In 2020, Petrosian and Pogozhev proposed an innovative solution method wherein players receive partial information over time, with the duration of this information being random. Lastly, Petrosian and Barabanov (2017) examined a forward-looking approach in cooperative differential games that involve uncertain stochastic dynamics.

This chapter extends subgame-consistent solutions to dynamic cooperative games characterized by a random horizon to fair gain-sharing solutions. We derive the random-horizon Bellman equation along with the Isaacs-Bellman equations. Furthermore, we establish subgame-consistent cooperative solutions for dynamic games with a random horizon. We also derive analytically tractable payoff distribution mechanisms that facilitate the realization of these solutions. This analysis broadens the applicability of cooperative dynamic game theory to scenarios where the game horizon is random.

Appendices

Appendix A. Proof of Proposition 5.1

Consider first the last stage, i.e., stage 4. Invoking that $\boldsymbol{V^i(4,x)} = [\boldsymbol{A_4^i x} + \boldsymbol{C_4^i}]$ from Proposition 5.1, $\boldsymbol{V^i(5,x)} = \boldsymbol{q^i x}\left(\frac{\boldsymbol{1}}{\boldsymbol{1+r}}\right)^{\boldsymbol{4}}$, and (41), the second equation in (39) becomes

$$
\begin{aligned}
&[A_4^i x + C_4^i] \\
&\quad = \left\{P\left[P - q^i\left(\frac{1}{1+r}\right)\right]\frac{x}{2c_i} - \left[P - q^i\left(\frac{1}{1+r}\right)\right]^2\frac{1}{4c_i}\right\}\left(\frac{1}{1+r}\right)^3 \\
&\qquad + q^i\left\{x + a - bx - \left[P - q^i\left(\frac{1}{1+r}\right)\right]\frac{x}{2c_i}\right. \\
&\qquad \left. - \left[P - q^j\left(\frac{1}{1+r}\right)\right]\frac{x}{2c_j}\right\}\left(\frac{1}{1+r}\right)^4. \qquad \text{(A.1)}
\end{aligned}
$$

Note that the expressions on both the left-hand side and right-hand side of equation (A.1) are linear in x, and one can readily obtain $\boldsymbol{A_4^i}$ and $\boldsymbol{C_4^i}$ explicitly as

$$
\begin{aligned}
A_4^i &= \left\{P\left[P - q^i\left(\frac{1}{1+r}\right)\right]\frac{1}{2c_i} - \left[P - q^i\left(\frac{1}{1+r}\right)\right]^2\frac{1}{4c_i}\right\}\left(\frac{1}{1+r}\right)^3 \\
&\quad + \boldsymbol{q^i}\left\{\boldsymbol{1} - \boldsymbol{b} - \left[\boldsymbol{P} - \boldsymbol{q^i}\left(\frac{\boldsymbol{1}}{\boldsymbol{1+r}}\right)\right]\frac{\boldsymbol{1}}{\boldsymbol{2c_i}}\right. \\
&\quad \left. - \left[\boldsymbol{P} - \boldsymbol{q^j}\left(\frac{\boldsymbol{1}}{\boldsymbol{1+r}}\right)\right]\frac{\boldsymbol{1}}{\boldsymbol{2c_j}}\right\}\left(\frac{\boldsymbol{1}}{\boldsymbol{1+r}}\right)^{\boldsymbol{4}}, \\
\boldsymbol{C_4^i} &= \boldsymbol{q^i a}. \qquad \text{(A.2)}
\end{aligned}
$$

Using (41) and Proposition 5.1, the game equilibrium strategy of player $\boldsymbol{i}$ in stage $\tau \in \{\boldsymbol{1,2,3}\}$ can be obtained as

$$
\begin{aligned}
\boldsymbol{u_\tau^{i**}} &= \left[\boldsymbol{P} - \frac{\boldsymbol{\theta_\tau}}{\sum_{\zeta=\tau}^{T}\boldsymbol{\theta_\zeta}}\boldsymbol{q^i}\left(\frac{\boldsymbol{1}}{\boldsymbol{1+r}}\right)\right. \\
&\quad \left. - \frac{\sum_{\zeta=\tau+1}^{T}\boldsymbol{\theta_\zeta}}{\sum_{\zeta=\tau}^{T}\boldsymbol{\theta_\zeta}}\boldsymbol{A_{x_{\tau+1}}^i}(\boldsymbol{1+r})^{\tau-1}\right]\frac{\boldsymbol{x}}{\boldsymbol{2c_i}}, \quad \boldsymbol{i} \in \{\boldsymbol{1,2}\}. \qquad \text{(A.3)}
\end{aligned}
$$

Once again invoking Proposition 5.1, the third set of equations in (39) can be expressed as

$$
\begin{aligned}
&[A^i_\tau x + C^i_\tau] \\
&= \Bigg\{ P \Bigg[P - \frac{\theta_\tau}{\sum_{\zeta=\tau}^{T}\theta_\zeta} q^i \left(\frac{1}{1+r}\right) \\
&\quad - \frac{\sum_{\zeta=\tau+1}^{T}\theta_\zeta}{\sum_{\zeta=\tau}^{T}\theta_\zeta} A^i_{x_{\tau+1}} (1+r)^{\tau-1} \Bigg] \frac{x}{2c_i} - \Bigg[P - \frac{\theta_\tau}{\sum_{\zeta=\tau}^{T}\theta_\zeta} q^i \left(\frac{1}{1+r}\right) \\
&\quad - \frac{\sum_{\zeta=\tau+1}^{T}\theta_\zeta}{\sum_{\zeta=\tau}^{T}\theta_\zeta} A^i_{x_{\tau+1}} (1+r)^{\tau-1} \Bigg]^2 \frac{x}{4c_i} \Bigg\} \left(\frac{1}{1+r}\right)^{\tau-1} \\
&\quad + \frac{\theta_\tau}{\sum_{\zeta=\tau}^{T}\theta_\zeta} q^i \Bigg\{ x + a - bx - \Bigg[P - \frac{\theta_\tau}{\sum_{\zeta=\tau}^{T}\theta_\zeta} q^i \left(\frac{1}{1+r}\right) \\
&\quad - \frac{\sum_{\zeta=\tau+1}^{T}\theta_\zeta}{\sum_{\zeta=\tau}^{T}\theta_\zeta} A^i_{x_{\tau+1}} (1+r)^{\tau-1} \Bigg] \frac{x}{2c_i} - \Bigg[P - \frac{\theta_\tau}{\sum_{\zeta=\tau}^{T}\theta_\zeta} q^j \left(\frac{1}{1+r}\right) \\
&\quad - \frac{\sum_{\zeta=\tau+1}^{T}\theta_\zeta}{\sum_{\zeta=\tau}^{T}\theta_\zeta} A^j_{x_{\tau+1}} (1+r)^{\tau-1} \Bigg] \frac{x}{2c_j} \Bigg\} \left(\frac{1}{1+r}\right)^{\tau} \\
&\quad + \frac{\sum_{\zeta=\tau+1}^{T}\theta_\zeta}{\sum_{\zeta=\tau}^{T}\theta_\zeta} \Bigg(A^i_{\tau+1} \Bigg\{ x + a - bx - \Bigg[P - \frac{\theta_\tau}{\sum_{\zeta=\tau}^{T}\theta_\zeta} q^i \left(\frac{1}{1+r}\right) \\
&\quad - \frac{\sum_{\zeta=\tau+1}^{T}\theta_\zeta}{\sum_{\zeta=\tau}^{T}\theta_\zeta} A^i_{x_{\tau+1}} (1+r)^{\tau-1} \Bigg] \frac{x}{2c_i} - \Bigg[P - \frac{\theta_\tau}{\sum_{\zeta=\tau}^{T}\theta_\zeta} q^j \left(\frac{1}{1+r}\right) \\
&\quad - \frac{\sum_{\zeta=\tau+1}^{T}\theta_\zeta}{\sum_{\zeta=\tau}^{T}\theta_\zeta} A^j_{x_{\tau+1}} (1+r)^{\tau-1} \Bigg] \frac{x}{2c_j} \Bigg\} + C^i_{\tau+1} \Bigg). \qquad \text{(A.4)}
\end{aligned}
$$

Once again, both the left-hand side and right-hand side of equation (A.4) are linear in $\boldsymbol{x}$, and one can readily obtain $\boldsymbol{A^i_\tau}$ and $\boldsymbol{C^i_\tau}$ in terms of the model parameters and $\boldsymbol{A^i_{\tau+1}}$ and $\boldsymbol{C^i_{\tau+1}}$.

Using $\boldsymbol{A^i_4}$ and $\boldsymbol{C^i_4}$ in (A.2), one can obtain $\boldsymbol{A^i_3}$ and $\boldsymbol{C^i_3}$ explicitly. Using $\boldsymbol{A^i_3}$ and $\boldsymbol{C^i_3}$, one can obtain $\boldsymbol{A^i_2}$ and $\boldsymbol{C^i_2}$ explicitly. Using $\boldsymbol{A^i_2}$ and $\boldsymbol{C^i_2}$, one can obtain $\boldsymbol{A^i_1}$ and $\boldsymbol{C^i_1}$ explicitly. Hence, Proposition 5.1 follows.

Appendix B. Proof of Proposition 5.2

Consider first the last stage, i.e., stage 4. Invoking that $\boldsymbol{W}(\mathbf{4},\boldsymbol{x}) = [\boldsymbol{A}_4\, \boldsymbol{x}+ \boldsymbol{C}_4]$ from Proposition 5.2, $\boldsymbol{W}(\mathbf{5},\boldsymbol{x}) = \sum_{j=1}^{2} q^j x \left(\frac{1}{1+r}\right)^4$, and (46), the second equation in (44) becomes

$$
\begin{aligned}
&[\boldsymbol{A}_4\, \boldsymbol{x}+ \boldsymbol{C}_4] \\
&= \sum_{j=1}^{2} \left\{ P \left[P - (q^1 + q^2) \left(\frac{1}{1+r} \right) \right] \frac{x}{2c_j} \right. \\
&\quad \left. - \left[P - (q^1 + q^2) \left(\frac{1}{1+r} \right) \right]^2 \frac{x}{4c_j} \right\} \left(\frac{1}{1+r} \right)^3 \\
&\quad + \sum_{j=1}^{2} q^j \left\{ x + a - bx - \left[P - (q^1 + q^2) \left(\frac{1}{1+r} \right) \right] \frac{x}{2c_1} \right. \\
&\quad \left. - \left[P - (q^1 + q^2) \left(\frac{1}{1+r} \right) \right] \frac{x}{2c_2} \right\} \left(\frac{1}{1+r} \right)^4 . \qquad \text{(B.1)}
\end{aligned}
$$

Note that the expressions on both the left-hand side and right-hand side of equation (B.1) are linear in x, and one can readily obtain $\boldsymbol{A}_4$ and $\boldsymbol{C}_4$ explicitly as

$$
\begin{aligned}
A_4 &= \sum_{j=1}^{2} \left\{ P \left[P - (q^1 + q^2) \left(\frac{1}{1+r} \right) \right] \frac{1}{2c_j} \right. \\
&\quad \left. - \left[P - (q^1 + q^2) \left(\frac{1}{1+r} \right) \right]^2 \frac{1}{4c_j} \right\} \left(\frac{1}{1+r} \right)^3 \\
&\quad + \sum_{j=1}^{2} \left\{ q^j \left\{ 1 - b - \left[P - (q^1 + q^2) \left(\frac{1}{1+r} \right) \right] \frac{1}{2c_1} \right. \right. \\
&\quad \left. - \left[P - (q^1 + q^2) \left(\frac{1}{1+r} \right) \right] \frac{1}{2c_2} \right\} \left(\frac{1}{1+r} \right)^4 , \\
\boldsymbol{C}_4 &= (\boldsymbol{q}^1 + \boldsymbol{q}^2)\boldsymbol{a}. \qquad \text{(B.2)}
\end{aligned}
$$

Using (46) and Proposition 5.2, the cooperative strategy of player $\boldsymbol{i}$ in stage $\boldsymbol{\tau \in \{1, 2, 3\}}$ can be obtained as

$$\boldsymbol{u_\tau^{i*}(x) = \left[P - \frac{\theta_\tau}{\sum_{\zeta=\tau}^{T}\theta_\zeta}(q^1 + q^2)\left(\frac{1}{1+r}\right) - \frac{\sum_{\zeta=\tau+1}^{T}\theta_\zeta}{\sum_{\zeta=\tau}^{T}\theta_\zeta}A_{x_{\tau+1}}(1+r)^{\tau-1}\right]\frac{x}{2c_i}} \tag{B.3}$$

for $\boldsymbol{i \in \{1, 2\}}$ and $\boldsymbol{\tau \in \{1, 2, 3\}}$.

Invoking Proposition 5.2, the third set of equations in (44) can be expressed as

$$\begin{aligned}
&[A_\tau x + C_\tau] \\
&= \sum_{j=1}^{n}\left\{P\left[P - \frac{\theta_\tau}{\sum_{\zeta=\tau}^{T}\theta_\zeta}(q^1 + q^2)\left(\frac{1}{1+r}\right)\right.\right. \\
&\quad \left. - \frac{\sum_{\zeta=\tau+1}^{T}\theta_\zeta}{\sum_{\zeta=\tau}^{T}\theta_\zeta}A_{x_{\tau+1}}(1+r)^{\tau-1}\right]\frac{x}{2c_j} \\
&\quad - \left[P - \frac{\theta_\tau}{\sum_{\zeta=\tau}^{T}\theta_\zeta}(q^1 + q^2)\left(\frac{1}{1+r}\right)\right. \\
&\quad \left.\left. - \frac{\sum_{\zeta=\tau+1}^{T}\theta_\zeta}{\sum_{\zeta=\tau}^{T}\theta_\zeta}A_{x_{\tau+1}}(1+r)^{\tau-1}\right]^2\frac{x}{4c_j}\right\}\left(\frac{1}{1+r}\right)^{\tau-1} \\
&\quad + \sum_{j=1}^{n}\frac{\theta_\tau}{\sum_{\zeta=\tau}^{T}\theta_\zeta}q^j\left\{x + a - bx - \left[P - \frac{\theta_\tau}{\sum_{\zeta=\tau}^{T}\theta_\zeta}(q^1 + q^2)\left(\frac{1}{1+r}\right)\right.\right. \\
&\quad \left. - \frac{\sum_{\zeta=\tau+1}^{T}\theta_\zeta}{\sum_{\zeta=\tau}^{T}\theta_\zeta}A_{x_{\tau+1}}(1+r)^{\tau-1}\right]\frac{x}{2c_1} \\
&\quad - \left[P - \frac{\theta_\tau}{\sum_{\zeta=\tau}^{T}\theta_\zeta}(q^1 + q^2)\left(\frac{1}{1+r}\right)\right.
\end{aligned}$$

$$
\begin{aligned}
&-\frac{\sum_{\zeta=\tau+1}^{T}\theta_\zeta}{\sum_{\zeta=\tau}^{T}\theta_\zeta}A_{x_{\tau+1}}(1+r)^{\tau-1}\Bigg]\frac{x}{2c_2}\Bigg\}\left(\frac{1}{1+r}\right)^{\tau}-\frac{\sum_{\zeta=\tau+1}^{T}\theta_\zeta}{\sum_{\zeta=\tau}^{T}\theta_\zeta}\\
&\times\left(A_{\tau+1}\Bigg\{x+a-bx-\Bigg[P-\frac{\theta_\tau}{\sum_{\zeta=\tau}^{T}\theta_\zeta}(q^1+q^2)\left(\frac{1}{1+r}\right)\right.\\
&-\frac{\sum_{\zeta=\tau+1}^{T}\theta_\zeta}{\sum_{\zeta=\tau}^{T}\theta_\zeta}A_{x_{\tau+1}}(1+r)^{\tau-1}\Bigg]\frac{x}{2c_1}\\
&-\Bigg[P-\frac{\theta_\tau}{\sum_{\zeta=\tau}^{T}\theta_\zeta}(q^1+q^2)\left(\frac{1}{1+r}\right)\\
&\left.-\frac{\sum_{\zeta=\tau+1}^{T}\theta_\zeta}{\sum_{\zeta=\tau}^{T}\theta_\zeta}A_{x_{\tau+1}}(1+r)^{\tau-1}\Bigg]\frac{x}{2c_2}\Bigg\}+C_{\tau+1}\right)
\end{aligned}
\tag{B.4}
$$

for $\tau \in \{1, 2, 3\}$.

Both the left-hand side and right-hand side of equation (B.4) are linear in $\boldsymbol{x}$, and one can readily obtain $\boldsymbol{A_\tau}$ and $\boldsymbol{C_\tau}$ in terms of the model parameters and $\boldsymbol{A_{\tau+1}}$ and $\boldsymbol{C_{\tau+1}}$.

Using $\boldsymbol{A_4}$ and $\boldsymbol{C_4}$ in (B.2), one can obtain $\boldsymbol{A_3}$ and $\boldsymbol{C_3}$ explicitly. Using $\boldsymbol{A_3}$ and $\boldsymbol{C_3}$, one can obtain $\boldsymbol{A_2}$ and $\boldsymbol{C_2}$ explicitly. Using $\boldsymbol{A_2}$ and $\boldsymbol{C_2}$, one can obtain $\boldsymbol{A_1}$ and $\boldsymbol{C_1}$ explicitly. Hence, Proposition 5.2 follows.

References

Ferreira, M., Pinheiro, D., & Pinheiro, S. (2019). Two-player zero-sum stochastic differential games with random horizon. *Advances in Applied Probability*, 51(4), 1209–1235.

Nash, J. F. (1951). Non-cooperative games: The Annals of Mathematics, v. 54.

Petrosian, O., & Barabanov, A. (2017). Looking forward approach in cooperative differential games with uncertain stochastic dynamics. *Journal of Optimization Theory and Applications*, 172(1), 328–347.

Petrosian, O., & Pogozhev, S. (2020). Looking forward approach with random horizon in cooperative differential games. *Contributions to Game Theory and Management*, 13, 360–387.

Petrosyan, L. A., & Mursov, N. V. (1966). Game theoretical problems in mechanics. *Lithuanian Mathematical Journal*, 6(3), 423–433.

Petrosyan, L. A., & Shevkoplyas, E. V. (2003). Cooperative solution for games with random duration. *Game Theory and Applications*, 9, 125–139.

Shevkoplyas, E. (2011). The Shapley value in cooperative differential games with random duration. In *Advances in Dynamic Games: Theory, Applications, and Numerical Methods for Differential and Stochastic Games* (pp. 359–373). Boston: Birkhäuser Boston.

Yeung, D. W. K., & Petrosyan, L. A. (2010). Subgame consistent solutions for cooperative stochastic dynamic games. *Journal of Optimization Theory and Applications*, 145(3), 579–596.

Yeung, D. W., & Petrosyan, L. A. (2011). Subgame consistent cooperative solution of dynamic games with random horizon. *Journal of Optimization Theory and Applications*, 150(1), 78–97.

Yeung, D. W. K., & Petrosyan, L. A. (2016). *Subgame Consistent Cooperation: A Comprehensive Treatise*, Singapore: Springer.

Zaremba, A., Gromova, E., & Tur, A. (2020). A differential game with random time horizon and discontinuous distribution. *Mathematics*, 8(12), 2185.

Chapter 6

Dynamic Fair Gain-Sharing in Games on a Network

"I have learned that a strong network is the ultimate shortcut to everything you want to achieve."
— **Meltta Campbell**

This chapter offers an analysis of the optimal management of interconnected networks. Collaboration is identified as the most effective means of achieving a collectively optimal solution that increases the payoffs for the players within a network. Given that players are not uniform, some may derive significantly less benefit from interconnection compared to their contributions to other providers. Therefore, an optimality principle of gain-sharing reflecting the contributions of the participating players is needed. This chapter introduces an innovative dynamic fair gain-sharing solution for interconnected networks that meets the criteria of individual rationality, group optimality, time consistency, and the fair-sharing principle. The organization of the chapter is as follows. Section 1 offers an overview of networks. Section 2 presents the formulation of an interconnected network, the non-cooperative equilibrium, efficient collaboration, and the marginal contributions of the players. In Section 3, a dynamic fair gain-sharing collaboration solution that reflects the contributions of the players is presented, and a time-consistent payoff distribution procedure (PDP) that leads to the achievement of the optimal collaboration solution is derived. An illustrative example of an interconnected electricity network is provided in Section 4. Finally, concluding remarks can be found in Section 5.

1. Networks

A network is a system composed of interconnected elements or participants, such as individuals, organizations, or nations, where the actions of these participants influence one another. This system enables the network participants to collaborate based on shared development goals outlined in a cooperation agreement. The players decide to join their strengths, share information, and create synergies to become more innovative and competitive while keeping their identity. Major types of networks include business networks, political networks, social networks, production networks, and international networks.

The optimal management of interconnected networks is exemplified by the establishment of a collaborative or cooperative framework. This framework has the potential to enhance investments in production capital, serve as a source for technology spillover, and create avenues for economic benefits. Given that interconnected networks function over time, a network dynamic game framework is one of the most efficient instruments to characterize the situation. Yeung *et al.* (2021, 2024a), Allen (2023), Pei *et al.* (2022, 2024), Tomassini and Pestelacci (2010), Grüner *et al.* (2013), Menache and Ozdaglar (2011), and Miles and Cavaliere (2022) have introduced network dynamic games. Furthermore, Petrosyan and Yeung (2021) along with Petrosyan *et al.* (2021) have examined network differential games. Mazalov and Chirkova (2019) offered an extensive analysis of the theory and applications of network games. A sustainable collaborative framework must exhibit the characteristic of time consistency. A collaborative solution is deemed time-consistent if the extension of the solution policy to a future time, starting from a later stage and a state resulting from previous optimal actions, remains optimal (refer to Yeung and Petrosyan, 2004, 2006, 2016; Yeung *et al.*, 2021, 2024b). Additionally, due to the asymmetry among players, some may realize significantly lower benefits from interconnection compared to their contributions to other providers. Therefore, it is essential to develop an optimality principle in gain-sharing that reflects the contributions of the participating players.

Two notable examples of networks are as follows:

(i) An interconnected electricity network serves as an electrical grid for the distribution of electricity from producers to consumers. These electrical grids can vary significantly in scale, potentially encompassing an entire nation or continent. They comprise power stations for electricity generation, electrical substations for voltage regulation, electric power transmission systems for long-distance power transport, and electric power distribution networks that deliver the necessary service voltages to individual customers.

(ii) A closed-loop supply chain employs reverse logistics to reintegrate products that have fulfilled their purpose back into the logistics framework. A manufacturer produces a product, dispatches it to customers, and subsequently encourages those customers to return the product when it is no longer needed. The returned items then re-enter the manufacturer's supply chain through processes, such as repair, resale, or component reuse.

2. A Dynamic Network Game

In this section, we formulate the dynamic interactions within a network as a dynamic network game. Let $N \equiv \{1, 2, \ldots, n\}$ be the set of players that are within a network. The planning horizon is T-stages. The set of players connected to player i is denoted by $K(i) = \{j : \text{arc}\,(i,j) \in L\}$ for $i \in N$. Note that $K(i)$ represents the technically possible set of connections available to player i. The network links are externally given, and a connection will be established when both firms agree to connect with each other. Any link will be cut off if one of the parties decides against connecting. Since there may exist a possibility that a certain player would receive zero gain from connecting with another player although it may give non-trivial gains to the players connected to it. Therefore, some players may not agree to be connected. Let $\bar{K}(i) \subseteq K(i)$ be the set of players that would mutually agree to maintain links with region i. The set $\bar{K}(i) = K(i)$ if all players in $K(i)$ would maintain links with firm i.

The set $\bar{K}(i)$ will be empty if all players in $K(i)$ do not maintain links with player i.

We use $x^i(t) \in X^i \subset R^{m_i}$ to denote the vector of state variables player i at stage t. The evolution of the state of player i is governed by

$$x_{t+1}^i = f_t^i(x_t^i, x_t^{K(i)}, u_t^i)$$
$$\text{for } x_1^i = x_1^{i(0)} \quad \text{for } i \in \{1, 2, \ldots, N\}, \tag{1}$$

where $x_t^{\bar{K}(i)}$ is the set of states of connected player j for $j \in \bar{K}(i)$ and $u_t^i \in U^i \subset R^{w_i}$ is player i's vector of strategies.

For notational convenience, we denote $x_t = (x_t^1, x_t^2, \ldots, x_t^n)$.

The payoff function of player i is

$$\sum_{t=1}^{T+1} \left[g_t^{i(i)}(x_t^i, u_t^i) + \sum_{j \in \bar{K}(i)} (g_t^{i(j)}(x_t^i, x_t^j, u_t^i, u_t^j)) \right] \delta_1^t$$
$$+ Q_{T+1}^i(x_{T+1}^i)\delta_1^{T+1} \quad \text{for } i \in N. \tag{2}$$

In particular, $g_t^{i(i)}(x_t^i, u_t^i) \geq 0$, $g_t^{i(j)}(x_t^j, x_t^i, u_t^i, u_t^j) \geq 0$, and $Q_{T+1}^i (x_{T+1}^i)\delta_1^{T+1} \geq 0$.

2.1. Efficient Collaboration and Pareto Optimum

Collaboration in the interconnected network suggests an efficient way of enhancing the joint payoff of all the players. We would first obtain the Pareto-optimal outcome. To ensure Pareto efficiency, the players in the network will maximize their joint payoffs. Since $g_t^{i(j)}(x_t^j, x_t^i, u_t^i, u_t^j) \geq 0$, all connections will be taken into consideration. The players have to solve the following dynamic optimization problem:

$$\max_{\substack{u_t^\ell, \ell \in N, \\ t \in \{1,2,\ldots,T\}}} \left\{ \sum_{\ell=1}^{n} \sum_{t=1}^{T+1} \left[g_t^{\ell(\ell)}(x_t^\ell, u_t^\ell) + \sum_{j \in K(\ell)} (g_t^{\ell(j)}(x_t^\ell, x_t^j, u_t^\ell, u_t^j)) \right] \delta_1^t \right.$$
$$\left. + \sum_{\ell=1}^{n} Q_{T+1}^\ell(x_{T+1}^\ell)\delta_1^{T+1} \right\} \tag{3}$$

subject to (1).

The maximized joint payoffs can be characterized by the following theorem.

Theorem 2.1. *A set of strategies* $\{u_t^{\ell*} = \psi_t^{\ell}(x)$ *for* $\ell \in N$ *and* $t \in \{1, 2, \ldots, T\}$ *constitutes an optimal solution to the optimization problems* (1) *and* (3) *if there exist functions* $W^N(t, x)$ *such that the following recursive relations are satisfied:*

$$W^N(T+1, x) = \sum_{\ell \in N} Q_{T+1}^{\ell}(x_{T+1}^{\ell})\delta_1^{T+1}, \tag{4}$$

$$\begin{aligned} W^N(t, x) = \max_{u_t^{\ell}, \ell \in N,} \Bigg\{ \sum_{\ell=1}^{n} \Bigg[& g_t^{\ell(\ell)}(x_t^{\ell}, u_t^{\ell}) \\ & + \sum_{j \in K(\ell)} (g_t^{\ell(j)}(x_t^{\ell}, x_t^{j}, u_t^{\ell}, u_t^{j})) \Bigg] \delta_1^t + W^N(t+1, x_{t+1}) \Bigg\} \\ & \text{for } t \in \{1, 2, \ldots, T\}, \end{aligned} \tag{5}$$

Proof. Equations (4)–(5) satisfy the optimal condition of the technique of dynamic programming, hence $W^N(t, x)$ is the maximized joint payoff of the n players. □

Substituting the optimal strategies $u_t^{\ell*}$ for $i \in N$ into (1), we obtain the dynamics of the state under cooperation as

$$x_{t+1}^i = f_t^i(x_t^i, x_t^{K(i)}, u_t^{i*}) \quad \text{for } i \in N. \tag{6}$$

We use $\{x_t^* \equiv \{x_t^{1*}, x_t^{2*}, \ldots, x_t^{n*}\}$ for $t \in \{1, 2, \ldots, T\}$ to denote the solution to (6), which is the optimal electricity-producing capitals under collaboration.

2.2. Marginal Contribution of Player in a Network

In order to distribute the cooperative gains from collaboration based on each player's contributions, we first need to assess the loss of joint payoff that occurs when a player withdraws from the collaborative arrangement. This loss can be determined by calculating the difference between the maximized joint payoff of the collaborative scheme with all players included and the maximized joint payoff when that specific player is excluded. If a player, say player i, cuts his connections with other members in the network, his link to all other players will be severed. To determine the maximized joint payoff without player i, we have to solve the game:

$$\max_{\substack{u_t^\ell, \ell \in N\backslash i,\\ t\in\{1,2,\ldots,T\}}} \left\{ \sum_{\ell \in N\backslash i} \sum_{t=1}^{T+1} \left[g_t^{\ell(\ell)}(x_t^\ell, u_t^\ell) + \sum_{\substack{j\in K(\ell)\\ j\neq i}} (g_t^{\ell(j)}(x_t^\ell, x_t^j, u_t^\ell, u_t^j)) \right] \delta_1^t \right.$$

$$\left. + \sum_{\ell\in N\backslash i} Q_{T+1}^\ell (x_{T+1}^\ell)\delta_1^{T+1} \right\}, \tag{7}$$

$$\max_{\substack{u_t^i\\ t\in\{1,2,\ldots,T\}}} \left\{ \sum_{t=1}^{T+1} \left[g_t^{i(i)}(x_t^i, u_t^i) + \sum_{j\in \bar{K}(i)} (g_t^{i(j)}(x_t^i, x_t^j, u_t^i, u_t^j)) \right] \delta_1^t \right.$$

$$\left. + Q_{T+1}^i (x_{T+1}^i)\delta_1^{T+1} \right\} \tag{8}$$

subject to

$$x_{t+1}^\ell = f_t^\ell(x_t^\ell, x_t^{K(\ell)\backslash i}, u_t^\ell) \quad \text{for } \ell \in N\backslash i, \quad \text{and}$$
$$x_{t+1}^i = f_t^i(x_t^i, u_t^i). \tag{9}$$

A feedback Nash equilibrium of the game (7)–(9) can be characterized by the following theorem.

Theorem 2.2. *A set of strategies* $\{\hat{u}_t^i = \hat{\varphi}_t^i(x), u_t^{(N\backslash i)\ell} = \varphi_t^{(N\backslash i)\ell}(x)\}$ *for* $\ell \in N\backslash i$ *and* $t \in \{1,2,\ldots,T\}$ *constitutes an equilibrium solution to the game (7)–(9) if there exist functions* $W^{N\backslash i}(t,x)$ *and* $\hat{V}^i(t,x)$ *such that the following recursive relations are satisfied:*

$$W^{N\backslash i}(T+1, x) = \sum_{\ell\in N\backslash i} Q_{T+1}^\ell (x_{T+1}^\ell)\delta_1^{T+1}, \tag{10}$$

$$W^{N\backslash i}(t,x)$$

$$= \max_{\substack{u_t^\ell \\ \ell\in N\backslash i}} \left\{ \sum_{\ell\in N\backslash i} \left[g_t^{\ell(\ell)}(x_t^\ell, u_t^\ell) + \sum_{\substack{j\in K(\ell) \\ j\neq i}} (g_t^{\ell(j)}(x_t^\ell, x_t^j, u_t^\ell, u_t^j)) \right] \delta_1^t \right.$$

$$\left. + W^{N\backslash i}(t+1, x_{t+1}^{(N\backslash i)}) \right\} \quad \text{for } t \in \{1,2,\ldots,T\}, \tag{11}$$

where $x_{t+1}^{(N\backslash i)}$ *is a vector with* $x_{t+1}^i = f_t^i(x_t^i, \hat{u}_t^i)$ *and* $x_{t+1}^\ell = f_t^\ell(x_t^\ell, x_t^{K(\ell)\backslash i}, u_t^\ell)$ *for* $\ell \in N\backslash i$,

$$\hat{V}^i(T+1,x) = Q_{T+1}^i(x_{T+1}^i)\delta_1^{T+1}, \tag{12}$$

$$\hat{V}^i(t,x) = \max_{u_t^i} \left\{ \left[g_t^{i(i)}(x_t^i, u_t^i) + \sum_{j\in \bar{K}(i)} (g_t^{i(j)}(x_t^i, x_t^j, u_t^i, u_t^j)) \right] \delta_1^t \right.$$

$$\left. + \hat{V}^i(t+1, \hat{x}_{t+1}) \right\} \quad \text{for } t \in \{1,2,\ldots,T\}, \tag{13}$$

where $\hat{x}_{t+1}$ *is a vector with* $x_{t+1}^i = f_t^i(x_t^i, u_t^i)$ *and* $x_{t+1}^\ell = f_t^\ell(x_t^\ell, x_t^{K(\ell)\backslash i}, u_t^{(N\backslash i)\ell})$ for $\ell \in N\backslash i$.

Proof. Invoking the technique of dynamic programming, $W^{N\backslash i}(t,x)$ is the maximized joint payoff of all nations except player i for the given Nash equilibrium strategies of player i and $\hat{V}^i(t,x)$ is the maximized payoff of player i for the given Nash equilibrium strategies of all other nations. Hence, a Nash equilibrium appears. □

3. Fair Gain-sharing Solutions and Payoff Distribution

This section first presents a fair gain-sharing solution for the above network game and then derives a time-consistent PDP so that the agreed-upon payoff will be realized. Again, consider the term

$$W^N(t,x) - W^{N\backslash i}(t,x) \quad \text{for } i \in N, \tag{14}$$

which yields the loss of payoffs when player i is not included in the network. It also reflects player i's contribution to the joint payoff. Any player wishing to drop out of the network will have all its links to other players, and its payoff will be $\hat{V}^i(t,x)$. The players in the network aim to share fairly the excess of the maximized cooperative payoff net of the sum of the players' drop-out payoffs $\hat{V}^j(t,x)$ for $j \in N$, i.e.,

$$\left[W^N(t,x) - \sum_{j=1}^{n} \hat{V}^j(t,x)\right]. \tag{15}$$

A fair measure of player i's contribution to the maximized joint payoff net of the total non-cooperative payoff is the weight

$$\frac{W^N(t,x) - W^{N\backslash i}(t,x)}{\sum_{\varsigma=1}^{n}\left[W^N(t,x) - W^{N\backslash \varsigma}(k,x,\vartheta)\right]} \quad \text{for player } i \in N. \tag{16}$$

The weight (16) reflects the proportion of player i's contribution to the total cooperative gain net of the total non-cooperative payoffs.

3.1. A Dynamic Fair Gain-sharing Solution

Using (15) and (16), a fair gain-sharing solution for the network game (1)–(2) can be obtained as

$$\begin{aligned}\xi^i(k,x_k) = {} & \frac{W^N(k,x_k) - W^{N\backslash i}(k,x_k)}{\sum_{\varsigma=1}^{n}\left[W^N(k,x_k) - W^{N\backslash \varsigma}(k,x_k)\right]} \\ & \times \left[W^N(k,x_k) - \sum_{j=1}^{n} \hat{V}^j(k,x_k)\right] + \hat{V}^i(k,x_k)\end{aligned} \tag{17}$$

for $i \in N$ at stage $k \in \{1,2,\ldots,T\}$ along the cooperative trajectory $\{x_k^*\}_{k=1}^T$.

The formula in (17) fulfills group optimality, individual rationality, time consistency, and the property of fair gain-sharing. It differs from the cooperative solution values in the previous chapters in that the network's non-cooperative payoff $\hat{V}^j(t,x)$ is not the same as $V^i(k,x_k^*)$.

3.2. Time-consistent Payoff Distribution Procedure

We have to formulate a PDP so that the agreed-upon payoff according to (17) will be realized. Following Yeung and Petrosyan (2010, 2016) and Yeung *et al.* (2024a, 2024b), we let $\beta_k^i(x_k^*)$ denote the payment that player i will receive at stage t. The payment scheme involving $\beta_k^i(x_k^*)$ constitutes a PDP of the payoff governed by (17) in that the payoff to player i from stage k to T can be expressed as

$$\xi^i(k,x_k) = \sum_{\tau=k}^{T} \beta_\tau^i(x_k)\delta_1^\tau + Q^i(x_{T+1}^{i*})\delta_1^{T+1}. \tag{18}$$

Theorem 3.1. *A payment $\beta^i(t,x^*)$ for $t \in \{1,2,\ldots,T\}$ equaling*

$$\beta_k^i(x_k^*) = (\delta_1^k)^{-1}[\xi^i(k,x_k^*) - \xi^i(k+1,x_{k+1}^*] \quad \text{for } i \in N \tag{19}$$

given to player i will lead to the realization of (17).

Proof. Using (23), we can obtain

$$\xi^i(k+1,x_{k+1}^*) = \sum_{\tau=k+1}^{T} \beta_\tau^i(x_t^*)\delta_1^k + Q^i(x_{T+1}^{i*})\delta_1^{T+1}. \tag{20}$$

Hence, we can express

$$\xi^i(k,x_k^*) = \beta_k^i(x_k^*)\delta_1^k + \psi^i(k+1,x_{k+1}^*). \tag{21}$$

Therefore,

$$\beta_k^i(x_k^*) = (\delta_1^k)^{-1}[\xi^i(k,x_k^*) - \xi^i(k+1,x_{k+1}^*] \quad \text{for } i \in N. \tag{22}$$

□

The PDP in (19) gives rise to the realization of the sharing of the cooperative payoff according to (17).

4. An Illustrative Example in Interconnected Electricity Network

Consider a network of interconnected electricity systems where various companies engage in distinct roles within the electric grid, such as electricity generation, power transmission, substations, and electricity distribution. The payoff function for the company $i \in N$ is

$$\sum_{t=1}^{T} \Bigg[p_t^{i(i)} (u_t^{i(i)})^{1/2} (x_t^i)^{1/2} - c_t^{i(i)} u_t^{i(i)} - w_t^i v_t^i$$
$$+ \sum_{j \in K(i)} (p_t^{i(j)} (u_t^{i(j)})^{1/2} (\varpi_t^{i(j)} x_t^j)^{1/2} + h_t^{i(j)} x_t^i - c_t^{i(j)} u_t^{i(j)}) \Bigg] \delta_1^t$$
$$+ (Q^i x_{T+1}^i + M^i) \delta_1^{T+1}$$
$$\text{for} \quad i \in N, \tag{23}$$

where $p_t^{i(i)} (u_t^{i(i)})^{1/2} (x_t^i)^{1/2}$ is the gross revenue of company i if it is not connected to other companies in the network, $c_t^{i(i)} u_t^{i(i)}$ is the input cost of $u_t^{i(i)}$, $w_t^i v_t^i$ is the cost of capital investment $v_t^{i(i)}$, $p_t^{i(j)} (u_t^{i(j)})^{1/2} (\varpi_t^{i(j)} x_t^j)^{1/2}$ is the gross revenue of company i can be obtained by linking with company $j \in K(i)i$, $c_t^{i(j)} u_t^{i(j)}$ is the input cost of $u_t^{i(j)}$, and $Q^i x_{T+1}^i + M^i$ is the terminal value of company i in stage $T+1$.

The dynamics of the electricity-generating capacity of company i is

$$x_{t+1}^i = x_t^i + a_t^i (v_t^i)^{1/2} + \sum_{j \in K(i)} \lambda_t^{i(j)} x_t^j - \sigma_t^i x_t^i \quad \text{for } i \in N, \tag{24}$$

where $\sum_{j \in K(i)} \lambda_t^{i(j)} x_t^j$ is the technology spill over from the capital of the connecting players.

4.1. Joint Payoff Maximization under Collaboration

We first consider the maximization of the joint payoff of all companies

$$\sum_{i\in N}\left\{\sum_{t=1}^{T}\left[p_t^{i(i)}(u_t^{i(i)})^{1/2}(x_t^i)^{1/2}-c_t^{i(i)}u_t^{i(i)}-w_t^iv_t^i\right.\right.$$
$$\left.+\sum_{j\in K(i)}(p_t^{i(j)}(u_t^{i(j)})^{1/2}(\varpi_t^{i(j)}x_t^j)^{1/2}-c_t^{i(j)}u_t^{i(j)})\right]\delta_1^t$$
$$\left.+(Q^ix_{T+1}^i+M^i)\delta_1^{T+1}\right\} \tag{25}$$

subject to dynamics (24).

Invoking Theorem 2.1, the Pareto-optimal joint payoff under collaboration can be obtained as follows.

Proposition 4.1.

$$W^N(t,x)=\left(\sum_{i\in N}A_t^{(N)i}x^i+C_t^N\right)\delta_t^t \quad \text{for } t\in\{1,2,\ldots,T\}, \tag{26}$$

where

$$A_{T+1}^{(N)i}=Q^i,\quad C_{T+1}^N=\sum_{i\in N}M^i;$$

$$A_t^{(N)i}=\left(\frac{(p_t^{i(i)})^2}{4c_t^{i(i)}}+\sum_{j\in K(i)}\frac{(p_t^{j(i)})^2}{4c_t^{j(i)}}\varpi_t^{j(i)}\right)+\delta_t^{t+1}A_{t+1}^{(N)i}(1-\sigma_t^i)$$
$$+\sum_{\varsigma\in K(i)}\delta_t^{t+1}A_{t+1}^{(N)\varsigma}\lambda_t^{\varsigma(i)},$$

$$C_t^N=-\sum_{i\in N}\frac{(a_t^i\delta_t^{T+1}A_{t+1}^{(N)i})^2}{4w_t^i}+\sum_{i\in N}\delta_t^{t+1}A_{t+1}^{(N)i}\left(a_t^i\frac{(a_t^i\delta_t^{t+1}A_{t+1}^{(N)i})}{2w_t^i}\right)$$
$$+\delta_t^{t+1}C_{t+1}^N$$
$$\text{for } t\in\{1,2,\ldots T\}. \tag{27}$$

Proof. See Appendix A. □

The optimal strategies under collaboration can be obtained as (see Appendix A)

$$
\begin{aligned}
u_t^{i(i)*} &= \left(\frac{p_t^{i(i)}}{2c_t^{i(i)}}\right)^2 x_t^i \quad \text{for } i \in N,\\
v_t^{i*} &= \frac{(a_t^i \delta_t^{t+1} A_t^{(N)i})^2}{4(w_t^i)^2} \quad \text{for } i \in N,\\
u_t^{i(j)*} &= \left(\frac{p_t^{i(j)}}{2c_t^{i(j)}}\right)^2 \varpi_t^{i(j)} x_t^j \quad \text{for } i \in N \quad \text{and} \quad j \in K(i).
\end{aligned}
\tag{28}
$$

The optimal collaborative state trajectory satisfies

$$
x_{t+1}^i = x_t^i + a_t^i \frac{a_t^i \delta_t^{t+1} A_t^{(N)i}}{2w_t^i} + \sum_{j \in K(i)} \lambda_t^{i(j)} x_t^j - \sigma_t^i x_t^i \quad \text{for } i \in N. \tag{29}
$$

We use $\{x_t^*\}_{t=1}^{T+1} = \{x_t^{1*}, x_t^{2*}, \ldots, x_t^{n*}\}_{t=1}^{T+1}$ to denote the solution to (29).

4.2. Maximized Joint Payoff without Company i

To obtain the maximized joint payoff without company i, we have to solve the game

$$
\max_{\substack{u_t^{\ell(\ell)}, u_t^{\ell(\varsigma)}, v_t^\ell \\ \text{for } \varsigma \in K(\ell)\backslash i, \\ \ell \in N\backslash i, \\ t \in \{1,2,\ldots,T\}}} \left\{ \sum_{\ell \in N\backslash i} \left\{ \sum_{t=1}^{T} \left[p_t^{\ell(\ell)} (u_t^{\ell(\ell)})^{1/2} (x_t^\ell)^{1/2} - c_t^{\ell(\ell)} u_t^{\ell(\ell)} - w_t^\ell v_t^\ell \right.\right.\right.
$$

$$
\left. + \sum_{\varsigma \in K(\ell)\backslash i} (p_t^{\ell(\varsigma)} (u_t^{\ell(\varsigma)})^{1/2} (\varpi_t^{\ell(\varsigma)} x_t^\varsigma)^{1/2} - c_t^{\ell(\varsigma)} u_t^{\ell(\varsigma)}) \right] \delta_1^t
$$

$$
\left.\left. + (Q^\ell x_{T+1}^\ell + M^\ell) \delta_1^{T+1} \right\}\right\},
$$

$$\max_{\substack{u_t^{i(i)},u_t^{i(\varsigma)},v_t^i \\ t\in\{1,2,\dots,T\}}} \left\{ \sum_{t=1}^{T} \left[p_t^{i(i)}(u_t^{i(i)})^{1/2}(x_t^i)^{1/2} - c_t^{i(i)}u_t^{i(i)} - w_t^i v_t^i \right] \delta_1^t \right.$$

$$\left. + (Q^i x_{T+1}^i + M^i)\delta_1^{T+1} \right\} \tag{30}$$

subject to dynamics

$$\begin{aligned} x_{t+1}^i &= x_t^i + a_t^i (v_t^i)^{1/2} - \sigma_t^i x_t^i, \\ x_{t+1}^\ell &= x_t^\ell + a_t^\ell (v_t^\ell)^{1/2} + \sum_{\varsigma\in K(\ell)\backslash i} \lambda_t^{\ell(\varsigma)} x_t^\varsigma - \sigma_t^\ell x_t^\ell \quad \text{for } \ell \in N\backslash i. \end{aligned} \tag{31}$$

Invoking Theorem 1.3, we can obtain the maximized joint payoff without company i independently as follows.

Proposition 4.2.

$$W^{N\backslash i}(t,x) = \left(\sum_{\ell\in N\backslash i} A_t^{(N\backslash i)\ell} x^\ell + C_t^{N\backslash i} \right) \delta_1^t \quad \text{for } t \in \{1,2,\dots,T\}, \tag{32}$$

where

$$\begin{aligned} A_t^{(N\backslash i)\ell} &= \left(\frac{(p_t^{\ell(\ell)})^2}{4c_t^{\ell(\ell)}} + \sum_{\varsigma\in K(\ell)\backslash i} \frac{(p_t^{\varsigma(\ell)})^2}{4c_t^{\varsigma(\ell)}} \varpi_t^{\varsigma(\ell)} \right) \\ &\quad + \delta_t^{t+1} A_{t+1}^{(N\backslash i)\ell} (1 - \sigma_t^{(N\backslash i)\ell}) + \sum_{\varsigma\in K(\ell)\backslash i} \delta_t^{t+1} A_{t+1}^{(N\backslash i)\varsigma} \lambda_t^{\varsigma(\ell)}, \\ C_t^{N\backslash i} &= - \sum_{\ell\in N\backslash i} \frac{(a_t^\ell \delta_t^{t+1} A_{t+1}^{(N\backslash i)\ell})^2}{4w_t^\ell} \\ &\quad + \sum_{\ell\in N\backslash i} \delta_t^{t+1} A_{t+1}^{(N\backslash i)\ell} \left(a_t^\ell \frac{(a_t^\ell \delta_t^{t+1} A_{t+1}^{(N\backslash i)\ell})}{2w_t^\ell} \right) + \delta_t^{t+1} C_{t+1}^{N\backslash i} \\ &\quad \text{for } t \in \{1,2,\dots,T\}. \end{aligned} \tag{33}$$

Proof. See Appendix B. □

The optimal strategies under collaboration in game (30)–(31) can be obtained as (see Appendix B)

$$\begin{aligned} u_t^{(N\backslash i)\ell(\ell)} &= \left(\frac{p_t^{\ell(\ell)}}{2c_t^{\ell(\ell)}}\right)^2 x_t^\ell \quad \text{for } \ell \in N\backslash i, \\ v_t^{(N\backslash i)\ell} &= \frac{(a_t^\ell \delta_t^{t+1} A_t^{(\ell)N\backslash i})^2}{4(w_t^\ell)^2} \quad \text{for } \ell \in N\backslash i, \\ u_t^{(N\backslash i)\ell(\varsigma)} &= \left(\frac{p_t^{\ell(\varsigma)}}{2c_t^{\ell(\varsigma)}}\right)^2 \varpi_t^{\ell(\varsigma)} \bar{x}_t^\ell \quad \text{for } \ell \in N\backslash i \quad \text{and} \quad \varsigma \in K(\ell)\backslash i. \end{aligned} \tag{34}$$

Invoking Theorem 3.1, we can obtain the payoff of company i when it acts independently as follows.

Proposition 4.3. *The payoff of company i when it acts independently is*

$$\hat{V}^i(t,x) = (\hat{A}_t^i x^i + \hat{C}_t^i)\delta_1^t \quad \text{for } t \in \{1,2,\ldots,T\}, \tag{35}$$

where

$$\begin{aligned} \hat{A}_t^i &= \frac{(p_t^{i(i)})^2}{4c_t^{i(i)}} + \delta_t^{t+1}\hat{A}_{t+1}^i(1-\sigma_t^i), \\ \hat{C}_t^i &= -\frac{(a_t^i\delta_t^{t+1}\hat{A}_{t+1}^i)^2}{4w_t^i} + \hat{A}_{t+1}^i\left(a_t^i\frac{(a_t^i\delta_t^{t+1}\hat{A}_{t+1}^i)}{2w_t^i}\right) + \delta_t^{t+1}\hat{C}_{t+1}^i. \end{aligned} \tag{36}$$

Proof. See Appendix B. □

The game equilibrium strategies of company i can be expressed as (See Appendix B)

$$\begin{aligned} \bar{u}_t^{i(i)} &= \left(\frac{p_t^{i(i)}}{2c_t^{i(i)}}\right)^2 x_t^i, \\ \hat{v}_t^i &= \frac{[a_t^i\delta_t^{t+1}\hat{A}_{t+1}^i]^2}{4(w_t^i)^2}. \end{aligned} \tag{37}$$

4.3. Dynamically Stable Fair Gain-sharing Solution and PDP

To obtain a fair gain-sharing solution, the companies consider the payoff $\hat{V}^i(t,x)$ as the non-cooperative payoff.

Invoking Formula 4.2, the payoff to company i under collaboration is

$$\begin{aligned}
\xi^i(t,x_t) &= \frac{W^N(t,x_t) - W^{N\backslash i}(t,x_t)}{\sum_{\varsigma=1}^{n}\left[W^N(t,x_t) - W^{N\backslash\varsigma}(t,x_t)\right]} \\
&\quad \times \left[W^N(t,x_t) - \sum_{j=1}^{n}\hat{V}^j(t,x_t)\right] + \hat{V}^i(t,x_t) \\
&= \frac{\sum_{j\in N} A_t^{(N)j}x_t^{j*} + C_t^N - \left(\sum_{\ell\in N\backslash i} A_t^{(N\backslash i)\ell}x_t^{\ell*} + C_t^{N\backslash i}\right)}{\sum_{\varsigma=1}^{n}\left[\sum_{j\in N} A_t^{(N)j}x_t^{j*} + C_t^N - \left(\sum_{\ell\in N\backslash\varsigma} A_t^{(N\backslash\varsigma)\ell}x_t^{\ell*} + C_t^{(N\backslash\varsigma)\ell}\right)\right]} \\
&\quad \times \left[\sum_{\ell\in N} A_t^{(N)\ell}x_t^{\ell*} + C_t^N - \sum_{j=1}^{n}(\hat{A}_t^j x_t^{j*} + \hat{C}_t^j)\right] + \hat{V}^i(t,x)
\end{aligned} \tag{38}$$

for $t \in \{1,2,\ldots,T\}$ along the cooperative trajectory $\{x_k^*\}_{k=1}^{T}$.

Invoking Theorem 3.1, the fair gain-sharing imputation $\xi(t,x_t^*)$ in (29) can be realized by a payment

$$\beta_t^i(x_t^*) = (\delta_1^t)^{-1}\left[\xi^i(t,x_t^*) - \xi^i(t+1,x_{t+1}^*)\right] \tag{39}$$

given to player $i \in N$ in stage $t \in \{1,2,\ldots,T\}$.

When all companies are using the collaborative strategies, the single-stage payoff that company i will receive directly at stage t becomes

$$\begin{aligned}
\pi_t^i(x_t^*) &= p_t^{i(i)}(u_t^{i(i)*})^{1/2}(x_t^{i*})^{1/2} - c_t^{i(i)}u_t^{i(i)*} - w_t^i v_t^i \\
&\quad + \sum_{j\in K(i)} p_t^{i(j)}(u_t^{i(j)})^{1/2}(\varpi_t^{i(j)}x_t^j)^{1/2} - c_t^{i(j)}u_t^{i(j)*}),
\end{aligned} \tag{40}$$

Therefore, an adjustment compensation $\Lambda_t^i(x_t^*) = \beta_t^i(x_t^*) - \pi^i(x_t^*)$ has to be given to company $i \in N$ at stage t.

5. Concluding Remarks

To effectively manage interconnected networks, collaboration emerges as the optimal approach to improve economic benefits, enhance production efficiency, and maximize the payoffs for the participants. Due to the asymmetrical nature of the players, some individuals may realize significantly lower gains from interconnection compared to their contributions to other players. This chapter introduces a dynamic fair gain-sharing solution for interconnected networks that adheres to the principles of individual rationality, group optimality, time consistency, and fair gain-sharing. Yeung *et al.* (2021, 2024a) explored network dynamic games within trade networks and the financial technology industry. Petrosyan and Yeung (2021) along with Petrosyan *et al.* (2021) examined network differential games. Mazalov and Chirkova (2019) offered an extensive analysis of the theory and applications related to network games. Kandogan (2008) derived consistent estimates regarding the trade impacts of regional blocs. Studies conducted by Eichengreen and Irwin (1995), Frankel and Rose (1998), McDonald *et al.* (2011), and Wolf and Ritschl (2011) focused on currency areas and trade blocs. Yeung and Lui (2019) investigated the political economy surrounding coordinated coalition cooperation, particularly in the context of China's Belt Road Initiative. Research on the interconnectivity of electricity networks can be referenced in studies (Purvins *et al.*, 2016; Meisen and Mohammadi, 2010; Zhang *et al.*, 2017; Yun and Zhang, 2006; Fan *et al.*, 2019; Rafique *et al.*, 2018).

Appendices

Appendix A. Proof of Proposition 4.1

Invoking Theorem 1.2, the solution to the maximization of the joint payoff under collaboration can be characterized by the system

$$W^N(T+1, x) = \sum_{i \in N} (Q^i x^i_{T+1} + M^i)\delta_1^{T+1}, \tag{A.1}$$

$$W^N(t,x) = \max_{\substack{u_t^{i(i)}, u_t^{i(j)}, v_t^i \\ \text{for } j \in K(i) \\ i \in N}} \left\{ \sum_{i \in N} \left[p_t^{i(i)} (u_t^{i(i)})^{1/2} (x_t^i)^{1/2} - c_t^{i(i)} u_t^{i(i)} - w_t^i v_t^i \right.\right.$$

$$\left. + \sum_{j \in K(i)} (p_t^{i(j)} (u_t^{i(j)})^{1/2} (\varpi_t^{i(j)} x_t^j)^{1/2} - c_t^{i(j)} u_t^{i(j)}) \right] \delta_1^t$$

$$\left. + W^N(t+1, x_{t+1}) \right\} \quad \text{for } t \in \{1, 2, \ldots, T\}, \tag{A.2}$$

where $x_{t+1} = (x_{t+1}^1, x_{t+1}^2, \ldots, x_{t+1}^n)$ with

$$x_{t+1}^i = x_t^i + a_t^i (v_t^i)^{1/2} + \sum_{j \in K(i)} \lambda_t^{i(j)} x_t^j - \sigma_t^i x_t^i \quad \text{for } i \in N.$$

Performing the indicated maximization yields optimal cooperative strategies

$$u_t^{i(i)} = \left(\frac{p_t^{i(i)}}{2c_t^{i(i)}} \right)^2 x_t^i \quad \text{for } i \in N,$$

$$v_t^i = \frac{[a_t^i (\delta_1^t)^{-1} \partial W^N(t+1, x_{t+1}) / \partial x_{t+1}^i]^2}{4(w_t^i)^2} \quad \text{for } i \in N,$$

$$u_t^{i(j)} = \left(\frac{p_t^{i(j)}}{2c_t^{i(j)}} \right)^2 \varpi_t^{i(j)} x_t^j \quad \text{for } i \in N \quad \text{and} \quad j \in K(i). \tag{A.3}$$

Substituting (A.3) into (A.2) yields

$$W^N(t,x) = \sum_{i \in N} \left[\frac{(p_t^{i(i)})^2}{4c_t^{i(i)}} x_t^i - \frac{[a_t^i (\delta_1^t)^{-1} \partial W(t+1, x_{t+1}) / \partial x_{t+1}^i]^2}{4w_t^i} \right.$$

$$\left. + \sum_{j \in K(i)} \left(\frac{(p_t^{i(j)})^2}{4c_t^{i(j)}} \varpi_t^{i(j)} x_t^j \right) \right] \delta_1^t + W^N(t+1, x_{t+1}), \tag{A.4}$$

where $x_{t+1} = (x_{t+1}^1, x_{t+1}^2, \ldots, x_{t+1}^n)$ with

$$x_{t+1}^i = x_t^i + a_t^i \frac{[a_t^i(\delta_1^t)^{-1}\partial W^N(t+1, x_{t+1})/\partial x_{t+1}^i]}{2w_t^i} + \sum_{j \in K(i)} \lambda_t^{i(j)} x_t^j - \sigma_t^i x_t^i \quad \text{for } i \in N.$$

We first start with the last stage of operation, i.e., stage T. With $W^N(T+1, x) = \sum_{i\in N}(Q^i x_{T+1}^i + M^i)\delta_1^{T+1} = \sum_{i\in N}(A_{T+1}^{(N)i} x^i + C_{T+1}^N)\delta_1^{T+1}$, we can obtain

$$v_T^i = \frac{[a_T^i(\delta_1^T)^{-1}\partial W^N(T+1, x_{t+1})/\partial x_{T+1}^i]^2}{4(w_T^i)^2} = \frac{[a_T^i \delta_T^{T+1} A_{T+1}^{(N)i}]^2}{4(w_T^i)^2}$$

for $i \in N$.

The stage T equation in (A.4) can be expressed as

$$\begin{aligned} W^N(T, x) = \sum_{i\in N} \Bigg[&\frac{(p_T^{i(i)})^2}{4c_T^{i(i)}} x_T^i - \frac{[a_T^i \delta_T^{T+1} A_{T+1}^{(N)i}]^2}{4w_T^i} \\ &+ \sum_{j\in K(i)} \left(\frac{(p_T^{i(j)})^2}{4c_T^{i(j)}} \varpi_T^{i(j)} x_T^j \right) \Bigg] \delta_1^t \\ &+ \Bigg[\sum_{i\in N} A_{T+1}^{(N)i} \Bigg(x_T^i + a_T^i \frac{[a_T^i \delta_T^{T+1} A_{T+1}^{(N)i}]}{2w_T^i} \\ &+ \sum_{j\in K(i)} \lambda_T^{i(j)} x_T^j - \sigma_T^i x_T^i \Bigg) + C_{T+1}^N \Bigg] \delta_1^{T+1}. \end{aligned} \tag{A.5}$$

Note that the right-hand side of (A.5) is a linear function of $x_T = (x_T^1, x_T^2, \ldots, x_T^n)$, therefore (A.5) can be written as

$$W^N(T, x) = \left(\sum_{i\in N} A_T^{(N)i} x^i + C_t^N \right) \delta_1^T, \tag{A.6}$$

where

$$A_T^{(N)i} = \left(\frac{(p_T^{i(i)})^2}{4c_T^{i(i)}} + \sum_{j\in K(i)} \frac{(p_T^{j(i)})^2}{4c_T^{j(i)}} \varpi_T^{j(i)} \right) + \delta_T^{T+1} A_{T+1}^{(N)i}(1-\sigma_T^i)$$
$$+ \sum_{\varsigma\in K(i)} \delta_T^{t+1} A_{T+1}^{(N)\varsigma} \lambda_T^{\varsigma(i)},$$

$$C_T^N = -\sum_{i\in N} \frac{(a_T^i \delta_T^{T+1} A_{T+1}^{(N)i})^2}{4w_T^i} + \sum_{i\in N} \delta_T^{t+1} A_{T+1}^{(N)i} \left(a_T^i \frac{(a_T^i \delta_T^{T+1} \hat{A}_{T+1}^{(N)i})}{2w_T^i} \right)$$
$$+ \delta_T^{T+1} C_{T+1}^N. \tag{A.7}$$

Repeating the analysis for $t \in \{T-1, T-2, \ldots, 1\}$, we can obtain

$$W^N(t,x) = \sum_{i\in N} \left[\frac{(p_t^{i(i)})^2}{4c_t^{i(i)}} x_t^i - \frac{[a_t^i \delta_t^{t+1} A_{t+1}^{(N)i}]^2}{4w_t^i} \right.$$
$$\left. + \sum_{j\in K(i)} \left(\frac{(p_t^{i(j)})^2}{4c_t^{i(j)}} \varpi_t^{i(j)} x_t^j \right) \right] \delta_1^t$$
$$+ \left[\sum_{i\in N} A_{t+1}^{(N)i} \left(x_t^i + a_t^i \frac{[a_t^i \delta_t^{T+1} A_{t+1}^{(N)i}]}{2w_t^i} \right. \right.$$
$$\left. \left. + \sum_{j\in K(i)} \lambda_t^{i(j)} x_t^j - \sigma_t^i x_t^i \right) + C_{t+1}^N \right] \delta_1^{t+1}. \tag{A.8}$$

Note that the right-hand side of (A.8) is a linear function of $x_{t+1} = (x_{t+1}^1, x_{t+1}^2, \ldots, x_{t+1}^n)$, therefore (A.8) can be written as

$$W^N(t,x) = \left(\sum_{i\in N} A_t^{(N)i} x^i + C_t^N \right) \delta_1^t, \tag{A.9}$$

where

$$A_t^{(N)i} = \left(\frac{(p_t^{i(i)})^2}{4c_t^{i(i)}} + \sum_{j\in K(i)} \frac{(p_t^{j(i)})^2}{4c_t^{j(i)}} \varpi_t^{j(i)} \right) + \delta_t^{T+1} A_{t+1}^{(N)i}(1-\sigma_t^i)$$
$$+ \sum_{\varsigma\in K(i)} \delta_t^{t+1} A_{t+1}^{(N)\varsigma} \lambda_t^{\varsigma(i)},$$

$$C_t^N = -\sum_{i\in N} \frac{(a_t^i \delta_t^{T+1} A_{t+1}^{(N)i})^2}{4w_t^i} + \sum_{i\in N} \delta_t^{t+1} A_{t+1}^{(N)i} \left(a_t^i \frac{(a_t^i \delta_t^{T+1} A_{t+1}^{(N)i})}{2w_t^i} \right)$$
$$+ \delta_t^{T+1} C_{t+1}^N. \tag{A.10}$$

The optimal strategies under collaboration can be expressed as

$$u_t^{i(i)} = \left(\frac{p_t^{i(i)}}{2c_t^{i(i)}} \right)^2 x_t^i \quad \text{for } i \in N,$$

$$v_t^i = \frac{(a_t^i \delta_t^{t+1} A_t^{(N)i})^2}{4(w_t^i)^2} \quad \text{for } i \in N,$$

$$u_t^{i(j)} = \left(\frac{p_t^{i(j)}}{2c_t^{i(j)}} \right)^2 \varpi_t^{i(j)} x_t^j \quad \text{for } i \in N \quad \text{and} \quad j \in K(i). \tag{A.11}$$

Hence, Proposition 4.1 follows.

Appendix B. Proof of Propositions 4.2 and 4.3

Invoking Theorem 1.3, the equilibrium solution to the game (30)–(31) can be characterized by the system

$$W^{N\backslash i}(T+1, x) = \sum_{\ell \in N\backslash i} (Q^\ell x_{T+1}^\ell + M^\ell) \delta_1^{T+1},$$

$$\hat{V}^i(T+1, x_{T+1}) = (Q^i x_{T+1}^i + M^i) \delta_1^{T+1}; \tag{B.1}$$

$$W^{N\backslash i}(t, x) = \max_{\substack{u_t^{\ell(\ell)}, u_t^{\ell(\varsigma)}, v_t^\ell \\ \text{for } \varsigma \in K(\ell)\backslash i, \\ \ell \in N\backslash i,}} \left\{ \sum_{\ell \in N\backslash i} \left[p_t^{\ell(\ell)} (u_t^{\ell(\ell)})^{1/2} (x_t^\ell)^{1/2} - c_t^{\ell(\ell)} u_t^{\ell(\ell)} - w_t^\ell v_t^\ell \right.\right.$$

$$\left. + \sum_{\varsigma \in K(\ell)\backslash i} (p_t^{\ell(\varsigma)} (u_t^{\ell(\varsigma)})^{1/2} (\varpi_t^{\ell(\varsigma)} x_t^\varsigma)^{1/2} - c_t^{\ell(\varsigma)} u_t^{\ell(\varsigma)}) \right] \delta_1^t$$

$$\left. + W^{N\backslash i}(t+1, x_{t+1}^{N\backslash i}) \right\} \quad \text{for } t \in \{1, 2, \ldots, T\}, \tag{B.2}$$

$$\hat{V}^i(t,x) = \max_{u_t^{i(i)},u_t^{i(\varsigma)},v_t^i} \left\{ \left[p_t^{i(i)}(u_t^{i(i)})^{1/2}(x_t^i)^{1/2} - c_t^{i(i)}u_t^{i(i)} - w_t^i v_t^i \right] \delta_1^t \right.$$
$$\left. + \hat{V}^i(t+1,\hat{x}_{t+1}) \right\} \quad \text{for } t \in \{1,2,\ldots,T\}, \tag{B.3}$$

where $x_{t+1}^{N\backslash i} = (x_{t+1}^1, x_{t+1}^2, \ldots, x_{t+1}^n)$ with

$$x_{t+1}^i = x_t^i + a_t^i(\hat{v}_t^i)^{1/2} - \sigma_t^i x_t^i,$$
$$x_{t+1}^\ell = x_t^\ell + a_t^\ell(v_t^\ell)^{1/2} + \sum_{\varsigma\in K(\ell)\backslash i} \lambda_t^{\ell(\varsigma)} x_t^\varsigma - \sigma_t^\ell x_t^\ell \quad \text{for } \ell \in N\backslash i;$$

$$\hat{x}_{t+1}^i = (x_{t+1}^1, x_{t+1}^2, \ldots, x_{t+1}^n) \text{ with}$$
$$x_{t+1}^i = x_t^i + a_t^i(v_t^i)^{1/2} - \sigma_t^i x_t^i,$$
$$x_{t+1}^\ell = x_t^\ell + a_t^\ell(v_t^{(N\backslash i)\ell})^{1/2} + \sum_{\varsigma\in K(\ell)\backslash i} \lambda_t^{\ell(\varsigma)} x_t^\varsigma - \sigma_t^\ell x_t^\ell \quad \text{for } \ell \in N\backslash i.$$

Performing the indicated maximization in (B.2)–(B.3) yields the equilibrium strategies:

$$u_t^{(N\backslash i)\ell(\ell)} = \left(\frac{p_t^{\ell(\ell)}}{2c_t^{\ell(\ell)}}\right)^2 x_t^\ell \quad \text{for } \ell \in N\backslash i,$$
$$v_t^{(N\backslash i)\ell} = \frac{[a_t^\ell(\delta_1^t)^{-1}\partial W^{N\backslash i}(t+1,x_{t+1})/\partial x_{t+1}^\ell]^2}{4(w_t^\ell)^2} \quad \text{for } \ell \in N\backslash i,$$
$$u_t^{(N\backslash i)\ell(\varsigma)} = \left(\frac{p_t^{\ell(\varsigma)}}{2c_t^{\ell(\varsigma)}}\right)^2 \varpi_t^{\ell(\varsigma)} x_t^\ell \quad \text{for } \ell \in N\backslash i \quad \text{and} \quad \varsigma \in K(\ell)\backslash i; \tag{B.4}$$

$$\hat{u}_t^{i(i)} = \left(\frac{p_t^{i(i)}}{2c_t^{i(i)}}\right)^2 x_t^i,$$
$$\hat{v}_t^i = \frac{[a_t^i(\delta_1^t)^{-1}\partial\hat{V}^i(t+1,x_{t+1})/\partial x_{t+1}^i]^2}{4(w_t^i)^2}. \tag{B.5}$$

Substituting (B.4)–(B.5) into (B.2)–(B.3) yields

$$W^{N\backslash i}(t,x) = \sum_{\ell \in N\backslash i} \left[\frac{(p_t^{\ell(\ell)})^2}{4c_t^{\ell(\ell)}} x_t^\ell - \frac{[a_t^\ell (\delta_1^t)^{-1} \partial W^{N\backslash i}(t+1, x_{t+1})/\partial x_{t+1}^\ell]^2}{4w_t^\ell} + \sum_{\varsigma \in K(\ell)\backslash i} \left(\frac{(p_t^{\ell(\varsigma)})^2}{4c_t^{\ell(\varsigma)}} \varpi_t^{\ell(\varsigma)} x_t^\varsigma \right) \right] \delta_1^t + W^{N\backslash i}(t+1, x_{t+1}^{N\backslash i}), \tag{B.6}$$

$$\hat{V}^i(t,x) = \left(\frac{(p_t^{i(i)})^2}{4c_t^{i(i)}} x_t^i - \frac{[a_t^i (\delta_1^t)^{-1} \partial \hat{V}^i(t+1, x_{t+1})/\partial x_{t+1}^i]^2}{4w_t^i} \right) \delta_1^t + \hat{V}^i(t+1, \hat{x}_{t+1}), \tag{B.7}$$

where $x_{t+1}^{N\backslash i} = (x_{t+1}^1, x_{t+1}^2, \ldots, x_{t+1}^n)$ with
$x_{t+1}^i = x_t^i + a_t^i \frac{[a_t^i (\delta_1^t)^{-1} \partial \hat{V}^i(t+1, x_{t+1})/\partial x_{t+1}^i]}{2w_t^i} - \sigma_t^i x_t^i$
$x_{t+1}^\ell = x_t^\ell + a_t^\ell (v_t^\ell)^{1/2} + \sum_{\varsigma \in K(\ell)\backslash i} \lambda_t^{\ell(\varsigma)} x_t^\varsigma - \sigma_t^\ell x_t^\ell$ for $\ell \in N\backslash i$;

$\hat{x}_{t+1}^i = (x_{t+1}^1, x_{t+1}^2, \ldots, x_{t+1}^n)$ with
$x_{t+1}^i = x_t^i + a_t^i (v_t^i)^{1/2} - \sigma_t^i x_t^i$,
$x_{t+1}^\ell = x_t^\ell + a_t^\ell \frac{[a_t^\ell (\delta_1^t)^{-1} \partial W^{N\backslash i}(t+1, x_{t+1})/\partial x_{t+1}^\ell]}{2w_t^\ell} + \sum_{\varsigma \in K(\ell)\backslash i} \lambda_t^{\ell(\varsigma)} x_t^\varsigma - \sigma_t^\ell x_t^\ell$ for $\ell \in N\backslash i$.

We first start with the last stage of operation, i.e., stage T. From (B.1), we obtain

$$W^{N\backslash i}(T+1, x) = \sum_{\ell \in N\backslash i} (Q^\ell x_{T+1}^\ell + M^\ell) \delta_1^{T+1} = \sum_{\ell \in N\backslash i} (A_{T+1}^{(N\backslash i)\ell} x^\ell + C_{T+1}^{N\backslash i}) \delta_1^{T+1},$$

$$V^i(T+1, x_{T+1}) = (Q^i x_{T+1}^i + M^i) \delta_1^{T+1} = (\hat{A}_{T+1}^i x^i + \hat{C}_{T+1}^i) \delta_1^{T+1}. \tag{B.8}$$

Note that $A_{T+1}^{(N\backslash i)i} = 0$ and $\hat{A}_{T+1}^\ell = 0$ for $\ell \in N\backslash i$.

We can obtain

$$v_T^{(N\backslash i)\ell} = \frac{[a_T^\ell(\delta_1^t)^{-1}\partial W^{N\backslash i}(T+1, x_{T+1})/\partial x_{T+1}^\ell]^2}{4(w_T^\ell)^2}$$

$$= \frac{[a_T^\ell \delta_T^{T+1} A_{T+1}^{(N\backslash i)\ell}]^2}{4(w_T^\ell)^2} \quad \text{for } \ell \in N\backslash i,$$

$$\hat{v}_T^i = \frac{[a_T^i(\delta_1^T)^{-1}\partial \hat{V}^i(T+1, x_{t+1})/\partial x_{T+1}^i]^2}{4(w_T^i)^2} = \frac{[a_T^i \delta_T^{T+1} \hat{A}_{T+1}^i]^2}{4(w_T^i)^2}. \tag{B.9}$$

The stage T equation in (B.6)–(B.7) can be expressed as

$$\begin{aligned} W^{N\backslash i}(T,x) = {} & \sum_{\ell \in N\backslash i}\left[\frac{(p_T^{\ell(\ell)})^2}{4c_T^{\ell(\ell)}}x_T^\ell - \frac{[a_T^\ell \delta_T^{T+1} A_{T+1}^{(N\backslash i)\ell}]^2}{4w_T^\ell}\right. \\ & \left. + \sum_{\varsigma \in K(\ell)\backslash i}\left(\frac{(p_T^{\ell(\varsigma)})^2}{4c_T^{\ell(\varsigma)}}\varpi_T^{\ell(\varsigma)}x_T^\varsigma\right)\right]\delta_1^T \\ & + \left[\sum_{\ell \in N\backslash i} A_{T+1}^{(N\backslash i)\ell}\left(x_T^\ell + a_T^\ell \frac{[a_T^\ell \delta_T^{T+1} A_{T+1}^{(N\backslash i)\ell}]}{2w_T^\ell}\right.\right. \\ & \left.\left. + \sum_{\varsigma \in K(\ell)\backslash i}\lambda_T^{l(\varsigma)}x_T^\varsigma - \sigma_T^\ell x_T^\ell\right) + C_{T+1}^{N\backslash i}\right]\delta_1^{T+1}, \end{aligned} \tag{B.10}$$

$$\begin{aligned} \hat{V}^i(T,x) = {} & \left(\frac{(p_T^{i(i)})^2}{4c_T^{i(i)}}x_T^i - \frac{[a_T^i \delta_T^{T+1}\hat{A}_{T+1}^i]^2}{4w_T^i}\right)\delta_1^T \\ & + \hat{A}_{T+1}^i\left(x_T^i + a_T^i\frac{[a_T^i\delta_T^{T+1}\hat{A}_{T+1}^i]}{2w_T^i}\right) + \hat{C}_{T+1}^i\Bigg]\delta_1^{T+1}. \end{aligned} \tag{B.11}$$

Note that the right-hand-side of equation system (B.10)–(B.11) are linear functions of $x_T = (x_T^1, x_T^2, \ldots, x_T^n)$, therefore we can obtain

$$W^{N\backslash i}(T,x) = \left(\sum_{\ell \in N\backslash i} A_T^{(N\backslash i)\ell}x^\ell + C_T^{N\backslash i}\right)\delta_1^T, \tag{B.12}$$

where

$$A_T^{(N\backslash i)\ell} = \left(\frac{(p_T^{\ell(\ell)})^2}{4c_T^{\ell(\ell)}} + \sum_{\varsigma \in K(\ell)\backslash i} \frac{(p_T^{\varsigma(\ell)})^2}{4c_T^{\varsigma(\ell)}} \varpi_T^{\varsigma(\ell)} \right) + \delta_T^{T+1} A_{T+1}^{(N\backslash i)\ell}(1 - \sigma_T^i)$$
$$+ \sum_{\varsigma \in K(\ell)\backslash i} \delta_T^{t+1} A_{T+1}^{(N\backslash i)\varsigma} \lambda_T^{\varsigma(\ell)},$$

$$C_T^{N\backslash i} = - \sum_{\ell \in N\backslash i} \frac{(a_T^\ell \delta_T^{T+1} A_{T+1}^{(N\backslash i)\ell})^2}{4w_T^\ell}$$
$$+ \sum_{\ell \in N\backslash i} \delta_T^{t+1} A_{T+1}^{(N\backslash i)\ell} \left(a_T^\ell \frac{[a_T^\ell \delta_T^{T+1} A_{T+1}^{(N\backslash i)\ell}]}{2w_T^\ell} \right) + \delta_T^{T+1} C_{T+1}^{N\backslash i},$$

$$\hat{V}^i(T, x) = (\hat{A}_T^i x^i + \hat{C}_T^i)\delta_1^T, \tag{B.13}$$

where

$$\hat{A}_T^i = \frac{(p_T^{i(i)})^2}{4c_T^{i(i)}} + \delta_T^{T+1} \hat{A}_{T+1}^i (1 - \sigma_T^i),$$

$$\hat{C}_T^i = -\frac{(a_T^i \delta_T^{T+1} \hat{A}_{T+1}^i)^2}{4w_T^i} + \hat{A}_{T+1}^i \left(a_T^i \frac{[a_T^i \delta_T^{T+1} \hat{A}_{T+1}^i]}{2w_T^i} \right) + \delta_T^{T+1} \hat{C}_{T+1}^i.$$

Note that $A_T^{(i)N\backslash i} = 0$ and $\hat{A}_T^\ell = 0$ for $\ell \in N\backslash i$.

Repeating the analysis for $t \in \{T-1, T-2, \ldots, 1\}$, we can obtain

$$W^{N\backslash i}(t, x) = \sum_{\ell \in N\backslash i} \left[\frac{(p_t^{\ell(\ell)})^2}{4c_t^{\ell(\ell)}} x_t^\ell - \frac{[a_t^\ell \delta_t^{t+1} A_{t+1}^{(N\backslash i)\ell}]^2}{4w_t^\ell} \right.$$
$$\left. + \sum_{\varsigma \in K(\ell)\backslash i} \left(\frac{(p_t^{\ell(\varsigma)})^2}{4c_t^{\ell(\varsigma)}} \varpi_t^{\ell(\varsigma)} x_t^\varsigma \right) \right] \delta_1^t$$
$$+ \left[\sum_{\ell \in N\backslash i} A_{t+1}^{(N\backslash i)\ell} \left(x_t^\ell + a_t^\ell \frac{[a_t^\ell \delta_t^{T+1} A_{t+1}^{(N\backslash i)\ell}]}{2w_t^\ell} \right. \right.$$

$$+ \sum_{\varsigma \in K(\ell)\backslash i} \lambda_t^{l(\varsigma)} x_t^{\varsigma} - \sigma_t^{\ell} x_t^{\ell} \Bigg) + C_{t+1}^{N\backslash i} \Bigg] \delta_1^{t+1}, \tag{B.14}$$

$$\hat{V}^i(t,x) = \left(\frac{(p_t^{i(i)})^2}{4c_t^{i(i)}} x_t^i - \frac{[a_t^i \delta_t^{t+1} \hat{A}_{t+1}^i]^2}{4w_t^i} \right) \delta_1^t$$
$$+ \hat{A}_{t+1}^i \left(x_t^i + a_t^i \frac{[a_t^i \delta_t^{t+1} \hat{A}_{t+1}^i]}{2w_t^i} + \sigma_t^i \right) + \hat{C}_{t+1}^i \Bigg] \delta_1^{t+1}. \tag{B.15}$$

Note that the right-hand-side of equation system (B.14)–(B.15) are linear functions of $x_t = (x_t^1, x_t^2, \ldots, x_t^n)$, therefore, we can obtain

$$W^{N\backslash i}(t,x) = \left(\sum_{\ell \in N\backslash i} A_t^{(N\backslash i)\ell} x^{\ell} + C_T^{N\backslash i} \right) \delta_1^t, \tag{B.16}$$

where

$$A_t^{(N\backslash i)\ell} = \left(\frac{(p_t^{\ell(\ell)})^2}{4c_t^{\ell(\ell)}} + \sum_{\varsigma \in K(\ell)\backslash i} \frac{(p_t^{\varsigma(\ell)})^2}{4c_t^{\varsigma(\ell)}} \varpi_t^{\varsigma(\ell)} \right)$$
$$+ \delta_t^{t+1} A_{t+1}^{(N\backslash i)\ell} (1 - \sigma_t^{(N\backslash i)\ell}) + \sum_{\varsigma \in K(\ell)\backslash i} \delta_t^{t+1} A_{t+1}^{(N\backslash i)\varsigma} \lambda_t^{\varsigma(\ell)},$$

$$C_t^{N\backslash i} = - \sum_{\ell \in N\backslash i} \frac{(a_t^{\ell} \delta_t^{t+1} A_{t+1}^{(N\backslash i)\ell})^2}{4w_t^{\ell}}$$
$$+ \sum_{\ell \in N\backslash i} \delta_t^{t+1} A_{t+1}^{(N\backslash i)\ell} \left(a_t^{\ell} \frac{[a_t^{\ell} \delta_t^{t+1} A_{t+1}^{(N\backslash i)\ell}]}{2w_t^{\ell}} \right) + \delta_t^{t+1} C_{t+1}^{N\backslash i}.$$

$$\hat{V}^i(t,x) = (\hat{A}_t^i x^i + \hat{C}_t^i) \delta_1^t, \tag{B.17}$$

where

$$\hat{A}_t^i = \frac{(p_t^{i(i)})^2}{4c_t^{i(i)}} + \delta_t^{t+1} \hat{A}_{t+1}^i (1 - \sigma_t^i),$$

$$\hat{C}_t^i = - \frac{(a_t^i \delta_t^{t+1} \hat{A}_{t+1}^i)^2}{4w_t^i} + \hat{A}_{t+1}^i \left(a_t^i \frac{(a_t^i \delta_t^{t+1} \hat{A}_{t+1}^i)}{2w_t^i} \right) + \delta_t^{t+1} \hat{C}_{t+1}^i.$$

The game equilibrium strategies can be expressed as

$$
\begin{aligned}
u_t^{(N\backslash i)\ell(\ell)} &= \left(\frac{p_t^{\ell(\ell)}}{2c_t^{\ell(\ell)}}\right)^2 x_t^\ell \quad \text{for } \ell \in N\backslash i,\\
v_t^{(N\backslash i)\ell} &= \frac{[a_t^\ell \delta_t^{t+1} A_{t+1}^{(N\backslash i)\ell}]^2}{4(w_t^\ell)^2} \quad \text{for } \ell \in N\backslash i,\\
\bar{u}_t^{\ell(\varsigma)N\backslash i} &= \left(\frac{p_t^{\ell(\varsigma)}}{2c_t^{\ell(\varsigma)}}\right)^2 \varpi_t^{\ell(\varsigma)} \bar{x}_t^\ell \quad \text{for } \ell \in N\backslash i \quad \text{and} \quad \varsigma \in K(\ell)\backslash i;\\
\bar{u}_t^{i(i)} &= \left(\frac{p_t^{i(i)}}{2c_t^{i(i)}}\right)^2 x_t^i,\\
\hat{v}_t^i &= \frac{[a_t^i \delta_t^{t+1} \hat{A}_{t+1}^i]^2}{4(w_t^i)^2}.
\end{aligned}
\tag{B.18}
$$

Hence, Propositions 6.2 and 6.3 follow.

References

Allen, B. (2023). Flipping the intuition for games on dynamic networks. *Nature Computational Science,* 3(9), 737–738. https://doi.org/10.1038/s43588-023-00513-3.

Eichengreen, B., & Irwin, D. A. (1995). Trade blocs, currency blocs and the reorientation of world trade in the 1930s. *Journal of International Economics*, 38(1–2), 1–24.

Fan, J., Wang, X., Huang, Q., Zhang, X., Li, Y., & Zeng, P. (2019). Power gird interconnection with HVDC link in Northeast Asia considering complementarity of renewable energy and time zone difference. *The Journal of Engineering*, 2019(16), 1625–1629.

Frankel, J. A., & Rose, A. K. (1998). The endogenity of the optimum currency area criteria. *The Economic Journal*, 108(449), 1009–1025.

Grüner, S., Radmacher, F. G., & Thomas, W. (2013). Connectivity games over dynamic networks. *Theoretical Computer Science*, 493, 46–65. https://doi.org/10.1016/j.tcs.2012.12.001.

International Political Economy (eds. Tony Yu and Dianna Kwan), Palgrave, pp. 197–226.

Kandogan, Y. (2008). Consistent estimates of regional blocs' trade effects. *Review of International Economics*, 16(2), 301–314.

Mazalov, V., & Chirkova, J. (2019). *Networking Games.* San Diego: Elsevier Science & Technology.

McDonald, F., Josef Tüselmann, H., Voronkova, S., & Golesorkhi, S. (2011). The strategic development of subsidiaries in regional trade blocs. *Multinational Business Review*, 19(3), 256–271.

Meisen, P., & Mohammadi, C. (2010). *Cross-Border Interconnections on Every Continent.* Global Energy Network Institute, 33.

Menache, I., & Ozdaglar, A. E. (2011). *Network Games: Theory, Models, and Dynamics.* Morgan & Claypool. ISBN-10: 1608454088.

Miles, A. L., & Cavaliere, M. (2022). Cooperation dynamics in dynamical networks with history-based decisions. *PloS One*, 17(11), e0275909–e0275909. https://doi.org/10.1371/journal.pone.0275909.

Pei, S., Cressman, R., & Zhang, B. (2022). Dynamic Games on Networks with Heterogeneous Players. Available at SSRN: https://ssrn.com/abstract=4217352 or http://dx.doi.org/10.2139/ssrn.4217352.

Pei, S., Cressman, R., & Zhang, B. (2024). Dynamic games on arbitrary networks with two types of players. *Journal of Mathematical Economics*, 113, 102990. https://doi.org/10.1016/j.jmateco.2024.102990.

Petrosyan, L. A., & Yeung, D. W. K. (2021). Shapley value for differential network games: Theory and application. *Journal of Dynamics & Games*, 8(2), 151–166. https://doi.org/10.3934/jdg.2020021.

Petrosyan, L. A., Yeung, D. W. K., & Pankratova, Y. (2021). Dynamic cooperative games on networks. In *International Conference on Mathematical Optimization Theory and Operations Research* (pp. 403–416). Cham: Springer International Publishing.

Purvins, A., Fulli, G., Covrig, C. F., Chaouachi, A., Bompard, E., Carpaneto, E., ... & Obushevs, A. (2016). *The Baltic Power System between East and West Interconnections.* Publications Office of the European Union: Luxembourg.

Rafique, S. F., Shen, P., Wang, Z., Rafique, R., Iqbal, T., Ijaz, S., & Javaid, U. (2018). Global power grid interconnection for sustainable growth: concept, project and research direction. *IET Generation, Transmission & Distribution*, 12(13), 3114–3123.

Tomassini, M., & Pestelacci, E. (2010). Coordination games on dynamical networks. *Games*, 1(3), 242–261. https://doi.org/10.3390/g1030242.

Wolf, N., & Ritschl, A. O. (2011). Endogeneity of currency areas and trade blocs: Evidence from a natural experiment. *Kyklos*, 64(2), 291–312.

Yeung, D. W. K., Petrosyan, L. A., & Zhang, Y. X. (2024a). A Dynamic Network Game of the Fintech Industry. *Journal of the Operations Research Society of China*, 12(1), 5–33.

Yeung, D. W. K. & Lui, A. W. C. (2019). China's Belt-Road Initiative: The Political Economy.

Yeung, D. W. K., & Petrosyan, L. A. (2004). Subgame consistent cooperative solutions in stochastic differential games. *Journal of Optimization Theory and Applications*, 120(3), 651–666.

Yeung, D. W. K., & Petrosyan, L. A. (2006). *Cooperative Stochastic Differential Games*. New York, NY: Springer.

Yeung, D. W. K., & Petrosyan, L. A. (2010). Subgame consistent solutions for cooperative stochastic dynamic games. *Journal of Optimization Theory and Applications*, 145(3), 579–596.

Yeung, D. W. K., & Petrosyan, L. A. (2016). Subgame consistent cooperation. A Comprehensive treatise. *Ser. Theory and Decision Library C*, 47.

Yeung, D. W. K., Petrosyan, L. A. & Zhang, Y. X. (2024b). A fair-sharing value for cooperative dynamic network games. *Dynamic Games and Applications*, https://doi.org/10.1007/s13235-024-00562-x.

Yeung, D. W. K., Petrosyan, L. A., & Zhang, Y. X. (2021). Trade with technology spillover: A dynamic network game analysis. *International Game Theory Review*, 23(01), 2050011.

Yun, W. C., & Zhang, Z. X. (2006). Electric power grid interconnection in Northeast Asia. *Energy Policy*, 34(15), 2298–2309.

Zhang, X. P., Ou, M., Song, Y., & Li, X. (2017). Review of Middle East energy interconnection development. *Journal of Modern Power Systems and Clean Energy*, 5(6), 917–935.

Chapter 7

Fair Gain-Sharing Under Dynamic Uncertainty

"Uncertainty is the only certainty there is, and knowing how to live with insecurity is the only security."

— **John Allen Paulos**

This chapter considers fair gain-sharing cooperative solutions in interactive situations with uncertain future conditions. The presence of random elements in future payoff structures and evolution dynamics is prevalent in many practical game situations like regional economic cooperation, corporate joint ventures, and transboundary environmental management. A gain-sharing model that reflects the contributions of the participants offers a fair solution for all involved amidst the unpredictable future changes. The chapter begins by formulating the scenario within a stochastic dynamic game framework. A fair and time-consistent gain-sharing solution is developed for the purpose of cooperation. An analytically manageable payoff distribution method, dependent on specific random realizations of the state and payoff structure, is established. Furthermore, the analysis is extended to encompass a system with uncertain future payoff structures and the random evolution of state variables. The chapter is structured as follows: Section 1 presents the game formulation, non-cooperative equilibria, and group optimal conditions in stochastic dynamic games. Section 2 addresses fair gain-sharing cooperation, deriving the marginal contributions, the fair gain-sharing weight, and the cooperative solution. An example of cooperative resource extraction under uncertainty is illustrated in Section 3. Section 4 presents a note on the uncertainties of the future payoff structures. Finally, concluding remarks are provided in Section 5.

1. Stochastic Dynamic Games

In this section, we consider the paradigm of stochastic dynamic games. We start with the formulation of a general class of stochastic dynamic games, followed by the derivation of the non-cooperative equilibrium and the group-optimal outcome. In stochastic dynamic games, the evolution dynamics of the state variables are subject to random shocks. It is not uncommon to observe that the changes of some state variables are influenced by random elements.

1.1. Examples of Stochastic Dynamics

Stochastic state dynamics exist in a wide range of fields, which include the following:

(i) Market and Financial Dynamics Conditions

- The accumulation of economic assets is sensitive to variations in interest rates, trade balances, and the values of stocks and shares, real estate, and commodities.
- The accumulation of physical capital due to random factors appearing in the conversion of investment into capital, as well as depreciation and obsolescence.
- The accumulation of intangible capital is subject to unpredictable changes in goodwill, consumer preferences, brand perception, social media influences, demand trends, and the effectiveness of advertising.
- The cycles of innovation involve breakthroughs and failures that often arise unexpectedly.
- Corporate reputation is susceptible to random incidents, whistle-blowing, or viral backlash.

(ii) Environmental Conditions with Random Evolution

- Air Pollution fluctuates according to unexpected changes in industrial activities, climatic conditions, and the enforcement of regulations.
- Water quality is affected by runoff, precipitation, and unintentional spills.
- Climate variability encompasses extreme weather phenomena, El Niño/La Niña patterns, and long-term changes that are not accurately predictable.
- Biodiversity loss is caused by habitat destruction, the introduction of invasive species, and random ecological disturbances.
- Soil degradation is influenced by unpredictable elements associated with land utilization and agricultural technology.
- Ocean acidification is affected by CO_2 concentrations, temperature, and biological responses.
- Deforestation can be intensified by droughts, wildfires, lightning strikes, and human carelessness and overexploitation.

Stochastic dynamic games manifest in numerous real-life situations where decisions evolve over time amidst uncertainty in state dynamics and involve multiple participants.

1.2. Game Formulation

Consider the T-stage n-player nonzero-sum dynamic game with initial state x^0. The state space of the game is $X \in R^m$ and the state dynamics of the game is characterized by the stochastic difference equation

$$x_{k+1} = f_k(x_k, u_k^1, u_k^2, \ldots, u_k^n) + \vartheta_k G(x_k) \quad \text{for} \quad k \in \{1, 2, \ldots, T\}$$
$$\text{and} \quad x_1 = x^0, \tag{1}$$

where $u_k^i \in U^i \subset R^{m_i}$ is the control vector of player i at stage k, $x_k \in X$ is the state, ϑ_k is a sequence of statistically independent random variables, and $G(x_k)$ is a differentiable function.

The expected payoff of player i at stage k is

$$E_{\vartheta_1,\vartheta_2,\ldots,\vartheta_T}\left\{\sum_{k=1}^{T} g_k^i(x_k, u_k^1, u_k^2, \ldots, u_k^n)\delta_1^k + q^i(x_{T+1})\delta_1^{T+1}\right\}$$

for $i \in \{1, 2, \ldots, n\} \equiv N$, (2)

where $E_{\vartheta_1,\vartheta_2,\ldots,\vartheta_T}$ is the expectation operation with respect to the random variables $\vartheta_1, \vartheta_2, \ldots, \vartheta_T$, and $q^i(x_{T+1})$ is a terminal payment given at stage $T+1$. The payoffs of the players are transferable.

1.3. Non-cooperative Equilibria

In this section, we characterize the non-cooperative outcome of the discrete-time economic game (9)–(10). A frequently used way to characterize and derive a feedback Nash equilibrium of the game is provided in the following theorem.

Theorem 1.1. *A set of strategies $\{u_k^{i**} = \varphi_k^i(x)$, for $k \in \{1, 2, \ldots, T\}$ and $i \in N\}$ provides a feedback Nash equilibrium solution to the game (2)–(3) if there exist functions $V^i(k, x)$, for $k \in \{1, 2, \ldots, T\}$ and $i \in N$, such that the following recursive relations are satisfied:*

$$V^i(T+1, x) = q_{T+1}^i(x)\delta_1^{T+1}; \tag{3}$$

$$\begin{aligned} V^i(k, x) = \max_{u_k^i} E_{\vartheta_k}\Big\{ & g_k^i(x, u_k^{1**}, u_k^{2**}, \ldots, u_k^{i-1**}, u_k^i, \\ & u_k^{i+1**}, \ldots, u_k^{n**})\delta_1^k + V^i[k+1, f_k(x, u_k^{1**}, u_k^{2**}, \ldots, \\ & u_k^{i-1**}, u_k^i, u_k^{i+1**}, \ldots, u_k^{n**}) + \vartheta_k G(x_k)]\Big\} \end{aligned} \tag{4}$$

for $i \in N$ and $k \in \{1, 2, \ldots, T\}$.

Proof. Invoking the discrete-time stochastic dynamic programming technique, $V^i(k, x)$ is the maximized payoff of player i for given strategies of the other $n-1$ players $\{u_k^{i**}(x)$, for $j \in N$ and $j \neq i\}$. Hence, a Nash equilibrium appears. □

1.4. Group Optimality

Now, consider the case when the players agree to cooperate and distribute the joint payoff among themselves according to an optimality principle. To achieve group optimality by maximizing their expected joint payoff, the players have to solve the discrete-time stochastic dynamic programming problem of maximizing

$$E_{\vartheta_1,\vartheta_2,\ldots,\vartheta_T}\left\{\sum_{j=1}^{n}\sum_{k=1}^{T}\left[g_k^j\left(x_k,u_k^1,u_k^2,\ldots,u_k^n\right)\delta_1^k\right]\right.$$
$$\left.+\sum_{j=1}^{n}q^j(x_{T+1})\delta_1^{T+1}\right\} \tag{5}$$

subject to (1).

The stochastic dynamic programming problem (1) and (5) can be regarded as a single decision-maker in the game with $n = 1$ and the payoff being the sum of all the players' payoffs.

Theorem 1.2. *A set of strategies* $\{u_k^{i*} = \varphi_k^i(x)$, *for* $k \in \{1, 2, \ldots, T\}$ *and* $i \in N\}$ *provides an optimal solution to the optimization problems (1) and (5) if there exist functions* $W^N(k, x)$, *for* $k \in \{1, 2, \ldots, T\}$, *such that the following recursive relations are satisfied:*

$$W^N(T+1,x) = \sum_{i=1}^{n} q_{T+1}^i(x)\delta_1^{T+1}, \tag{6}$$

$$W^N(k,x) = \max_{u_k^1,u_k^2,\ldots,u_k^n} E_{\vartheta_k}\left\{\sum_{i=1}^{n} g_k^i(x,u_k^1,u_k^2,\ldots,u_k^n)\delta_1^k\right.$$
$$\left.+W^N[k+1,f_k(x,u_k^1,u_k^2,\ldots,u_k^n)+\vartheta_k G(x_k)]\right\} \tag{7}$$

for $k \in \{1, 2, \ldots, T\}$.

Proof. Invoking the discrete-time stochastic dynamic programming technique, $W^N(k, x)$ is the maximized payoff of the optimization problems (1) and (5). □

Substituting the optimal strategies $\{u_k^{i*} = \varphi_k^i(x)$ for $k \in \{1, 2, \ldots, T\}$ and $i \in N\}$ into dynamics (1) yields the optimal evolution dynamics of the state:

$$x_{k+1} = f_k(x_k, u_k^{1*}, u_k^{2*}, \ldots, u_k^{n*}) + \vartheta_k G(x_k), \quad x_1 = x^0. \tag{8}$$

We use X_k^* to denote the set of realizable values of x_k^* at stage k generated by (8). The term $x_k^* \in X_k^*$ is used to denote an element in X_k^*.

2. Fair Gain-Sharing Cooperation

The presence of asymmetric participants in a cooperative stochastic dynamic game results in varying contributions to the maximized joint payoff. Certain players may receive significantly less benefit compared to their contributions to the cooperative efforts of others. Due to the asymmetry in the contributions made and the gains received, it is essential to implement a fair distribution scheme that reflects the players' contributions amidst uncertain future outcomes. Initially, we compute the expected marginal contribution of a player to the overall joint payoff. Subsequently, we present a fair gain-sharing weight that reflects the contributions of the players within the cooperative framework. Finally, a subgame-consistent fair gain-sharing solution is derived with a corresponding payment mechanism.

2.1. Marginal Contributions of Players

To derive a dynamic fair gain-sharing cooperative solution, we first calculate the additional value to the joint payoff of cooperating players if an additional player, say player i, is included in the cooperation pack. For notational convenience, we use Ni to denote the set of players without player i. The addition to the maximized joint payoff by including player i is the difference between the maximized expected joint payoff and the maximized expected joint payoff without player i. Since the addition to the expected maximized payoff by player i is the difference between the expected maximized joint payoff of the collaborating scheme and the expected maximized payoff without i, we have to solve the stochastic dynamic game:

$$\max_{u_T^1, u_T^2, \ldots, u_T^n} E_{\vartheta_1, \vartheta_2, \ldots, \vartheta_T} \left\{ \sum_{j \in N \backslash i} \sum_{k=1}^{T} \left[g_k^\ell \left(x_k, u_k^1, u_k^2, \ldots, u_k^n \right) \delta_1^k \right] + \sum_{\ell \in N \backslash i} q^\ell \left(x_{T+1} \right) \delta_1^{T+1} \right\}, \tag{9}$$

$$\max_{u_T^i} E_{\vartheta_1, \vartheta_2, \ldots, \vartheta_T} \sum_{k=1}^{T} g_k^i \left(x_k, u_k^1, u_k^2, \ldots, u_k^n \right) \delta_1^k + q^i \left(x_{T+1} \right) \delta_1^{T+1} \Big\} \tag{10}$$

subject to dynamics (1).

The equilibrium of the game (1) and (9)–(10) can be characterized by the following theorem.

Theorem 2.1. *A set of strategies* $\{\hat{u}_k^i = \hat{\varphi}_k^i(x)$ *and* $u_k^{(N\backslash i)\ell} = \varphi_k^{(N\backslash i)\ell}(x)$ *for* $\ell \in N\backslash i$ *and* $k \in \{1, 2, \ldots, T\}\}$ *provides an equilibrium to the game (1) and (9)–(10) if there exist functions* $\hat{V}^i(k, x)$ *and* $W^{N\backslash i}(k, x)$ *for* $k \in \{1, 2, \ldots, T\}$, *such that the following recursive relations are satisfied:*

$$W^{N\backslash i}(T+1, x) = \sum_{\ell \in N\backslash i} q_{T+1}^i(x) \delta_1^{T+1}; \tag{11}$$

$$W^{N\backslash i}(k, x) = \max_{u_k^\ell, \text{for} \ell \in N\backslash i} E_{\vartheta_k} \left\{ \sum_{\ell \in N\backslash i} g_k^\ell(x, u_k^1, u_k^2, \ldots, u_k^n) \delta_1^k + W^{N\backslash i}[k+1, f_k(x, u_k^1, u_k^2, \ldots, u_k^{i-1}, \hat{u}_k^i, \hat{u}_k^{i+1}, \ldots, u_k^n) + \vartheta_k G(x_k)] \right\} \tag{12}$$

for $k \in \{1, 2, \ldots, T\}$.

$$\widehat{V}^i(T+1,x) = q^i_{T+1}(x)\delta_1^{T+1}; \tag{13}$$

$$\begin{aligned}&\widehat{V}^i(k,x)\\ &= \max_{u^i_k} E_{\vartheta_k}\Big\{g^i_k(x, u_k^{(N\backslash i)1}, u_k^{(N\backslash i)2}, \ldots, u_k^{(N\backslash i)i-1}, u^i_k,\\ &\quad u_k^{(N\backslash i)i+1}, \ldots, u_k^{(N\backslash i)n})\delta_1^k + V^i[k+1, f_k(x, u_k^{(N\backslash i)1}, u_k^{(N\backslash i)2},\\ &\quad \ldots, u_k^{(N\backslash i)i-1}, u^i_k, u_k^{(N\backslash i)i+1}, \ldots, u_k^{(N\backslash i)n})\delta_1^k + \vartheta_k G(x_k)]\Big\}\end{aligned} \tag{14}$$

for $k \in \{1, 2, \ldots, T\}$.

Proof. Invoking the discrete-time stochastic dynamic programming technique, $\widehat{V}^i(k,x)$ is the maximized payoff of player i given the strategies $u_k^{(N\backslash i)l}$ for $l \in N\backslash i$ and $W^{N\backslash i}(k,x)$ is the maximized payoff of the coalition of $N\backslash i$ players given the strategies $\hat{u}^i_k$ of player i. Hence, a Nash equilibrium appears.

The characterization of the game equilibrium in Theorem 2.1 allows one to derive the expected joint payoff of players if player i is excluded. □

2.2. Fair Gain-Sharing Weight and Cooperative Solution

In a cooperation scheme, the participating players would share the excess of the maximized cooperative joint payoff over the sum of the non-cooperative payoff, i.e.,

$$\left[W^N(k,x) - \sum_{j=1}^{n} V^j(k,x)\right]. \tag{15}$$

A sustainable sharing mechanism has to reflect the players' contributions in the cooperative scheme. To resolve the problem, we develop a gain-sharing weight that reflects the participating players' contributions. Note that the addition to the maximized joint payoff by including player i is the difference between the maximized joint payoff and the maximized joint payoff without player i, i.e.,

$$W^N(k,x) - W^{N\backslash i}(k,x) \quad \text{for } i \in N. \tag{16}$$

To measure player i's share of contribution to the maximized joint payoff when all players' marginal contributions are taken into consideration, we adopt the weight

$$\frac{W^N(k,x) - W^{N\backslash i}(k,x)}{\sum_{\varsigma=1}^{n}\left[W^N(k,x) - W^{N\backslash \varsigma}(k,x)\right]} \quad \text{for player } i \in N. \tag{17}$$

The weight (18) reflects the relative size of player i's contribution to the cooperative gain.

A dynamically stable fair gain-sharing value can be obtained as

$$\xi^i(k,x_k^*) = \frac{W^N(k,x_k^*) - W^{N\backslash i}(k,x_k^*)}{\sum_{\varsigma=1}^{n}\left[W^N(k,x_k^*) - W^{N\backslash \varsigma}(k,x_k^*)\right]} \times \left[W^N(k,x_k^*) - \sum_{j=1}^{n} V^j(k,x_k^*)\right] + V^i(k,x_k^*) \tag{18}$$

for $i \in N$ along the optimal cooperative trajectory $\{x_k^*\}_{k=1}^T$.

2.3. Subgame Consistent Solutions and Payment Mechanism

Following the analysis of Yeung and Petrosyan (2010, 2011), we formulate a payoff distribution procedure (PDP) so that the agreed-upon imputation (18) can be realized. Let $B_k^i(x_k^*)$ denote the payment that player i will receive at stage k under the cooperative agreement in (18) if $x_k^* \in X_k^*$ is realized. The payment scheme $B_k^i(x_k^*)$ constitutes a PDP in the sense that the imputation to player i over the stages k to $T+1$ can be expressed as

$$\xi^i(k,x_k^*) = B_k^i(x_k^*)\,\delta_1^k + E_{\vartheta_{k+1},\vartheta_{k+2},\ldots\ldots,\vartheta_T} \times \left(\sum_{\zeta=k+1}^{T} B_\zeta^i(x_\zeta^*)\,\delta_1^\xi + q^i(x_{T+1}^*)\,\delta_1^{T+1}\right) \tag{19}$$

for $i \in N$.

A theorem for the derivation of a subgame-consistent PDP can be established as follows.

Theorem 2.2. *A payment equaling*

$$B_k^i(x_k^*) = \frac{1}{\delta_1^k}\left[\xi^i(k, x_k^*) - E_{\vartheta_{k+1}}(\xi^i[k+1, f_k(x_k^*, u_k^{1*}, u_k^{2*}, \ldots, u_k^{n*}) + \vartheta_k G_k(x_k^*)])\right] \tag{20}$$

given to player $i \in N$ at stage $k \in \{1, 2, \ldots, T\}$ if $\theta_k^{\sigma_k}$ occurs and $x_k^* \in X_k^*$ leads to the realization of the imputation in (18).

Proof. Using (19), we obtain

$$\xi^i(k+1, x_k^*) = B_{k+1}^i(x_{k+1}^*)\delta_1^{k+1} + E_{\vartheta_{k+2},\vartheta_{k+3},\ldots,\vartheta_T}\left(\sum_{\zeta=k+2}^{T} B_\zeta^i(x_\zeta^*)\delta_1^\zeta + q^i(x_{T+1}^*)\delta_1^{T+1}\right). \tag{21}$$

Substituting (21) into (19) yields

$$\xi^i(k, x_k^*) = B_k^i(x_k^*)\delta_1^k + E_{\vartheta_k+1}(\xi^i[k+1, f_k(x_k^*, u_k^{1*}, u_k^{2*}, \ldots, u_k^{n*}) + \vartheta_k G_k(x_k^*)]). \tag{22}$$

Hence,

$$B_k^i(x_k^*) = \frac{1}{\delta_1^k}\left[\xi^i(k, x_k^*) - E_{\vartheta_{k+1}}\left(\xi^i\left[k+1, f_k\left(x_k^*, u_k^{1*}, u_k^{2*}, \ldots, u_k^{n*}\right) + \vartheta_k G_k\left(x_k^*\right)\right]\right)\right]. \tag{23}$$

□

When all players are using the cooperative strategies, the payoff that player i will directly receive at stage k given that $x_k^* \in X_k^*$ is

$$g_k^i(x_k^*, u_k^{1*}, u_k^{2*}, \ldots, u_k^{n*}).$$

However, according to the agreed-upon imputation, player i is supposed to receive $B_k^i(x_k^*)$ at stage k, as given in Theorem 2.2.

Therefore, a transfer payment (which can be positive or negative)

$$\varpi_k^i(x_k^*) = B_k^i(x_k^*) - g_k^i(x_k^*, u_k^{1*}, u_k^{2*}, \ldots, u_k^{n*})$$
$$\text{for } k \in \{1, 2, \ldots, T\} \quad \text{and } i \in N \tag{24}$$

will be assigned to player i.

The transfer payments system in (24) constitutes an instrument to guide the execution of the agreed-upon payoff sharing mechanism.

3. An Illustration in Cooperative Resource Extraction

Consider a T-stage variant of the stochastic dynamic resource extraction game in Yeung and Petrosyan (2012), where there are two resource extractors (firms). The lease for resource extraction begins at stage 1 and ends at stage T. Let u_k^i denote the resource extracted by firm i at stage k for $i \in \{1, 2\}$. Let U^i be the set of admissible amounts of resource extracted by firm i and $x_k \in X \subset R^+$ be the size of the resource stock at stage k.

The payoffs firm 1 and firm 2 will obtain at stage k are, respectively, as follows:

$$\left[\boldsymbol{P_k^1 u_k^1} - \frac{\boldsymbol{c_k^1}}{\boldsymbol{x_k}}\left(\boldsymbol{u_k^1}\right)^{\boldsymbol{2}}\right]\boldsymbol{\delta_1^k}, \quad \text{and}$$
$$\left[\boldsymbol{P_k^2 u_k^2} - \frac{\boldsymbol{c_k^2}}{\boldsymbol{x_k}}\left(\boldsymbol{u_k^2}\right)^{\boldsymbol{2}}\right]\boldsymbol{\delta_1^k}, \quad \text{for} \quad \boldsymbol{k} \in \{\boldsymbol{1}, \boldsymbol{2}, \ldots, \boldsymbol{T}\}, \tag{25}$$

where P_k^i is the price of the resource extracted and processed by firm i and $c_k^i(u_k^i)^2/x_k$ is the production cost of firm i in stage k.

In stage $T+1$, a terminal payment contingent upon the resource size equaling $q_{T+1}^i x_{T+1}$ will be paid to firm i.

The growth dynamics of the resource is governed by the stochastic difference equation

$$x_{k+1} = x_k + a - bx_k - \sum_{j=1}^{2} u_k^j + \vartheta_k x_k \quad \text{for} \quad k \in \{1, 2, \ldots, T\}$$
$$\text{and} \quad x_1 = x^0, \tag{26}$$

where ϑ_k is a random variable with non-negative range $\{\vartheta_k^1, \vartheta_k^2, \vartheta_k^3, \vartheta_k^4\}$ and corresponding probabilities $\{\gamma_k^1, \gamma_k^2, \gamma_k^3, \gamma_k^4\}$;

moreover, ϑ_1, $\vartheta_2, \ldots, \vartheta_T$ are independent. In addition, we have the constraint $u_k^1 + u_k^2 \leq (1-b)x_k + a$.

The objective of extractor $i \in \{1, 2\}$ is to maximize the present value of the expected stream of future profits:

$$\boldsymbol{E}_{\vartheta_1,\vartheta_2,\ldots,\vartheta_T}\left\{\sum_{k=1}^{T}\left[P_k^i u_k^i - \frac{c_k^i}{x_k}(u_k^i)^2\right]\delta_1^k \right.$$

$$\left. + (q_{T+1}^i x_{T+1} + Q_{T+1}^i)\delta_1^{T+1}\right\} \tag{27}$$

subject to the stochastic dynamics (26).

Invoking Theorem 1.1, one can characterize the non-cooperative Nash equilibrium strategies for the game (26)–(27) in the following corollary.

Corollary 3.1. *A set of strategies* $\{u_k^{i**} = \varphi_k^i(x)$ *for* $k \in \{1, 2, \ldots, T\}$ *and* $i \in \{1, 2\}\}$ *provides a feedback Nash equilibrium solution to the game (26)–(27) if there exist functions* $V^i(k, x)$, *for* $k \in \{1, 2, \ldots, T\}$ *and* $i \in N$, *such that the following recursive relations are satisfied:*

$$V^i(T+1, x) = (q_{T+1}^i x_{T+1} + Q_{T+1}^i)\delta_1^{T+1}, \tag{28}$$

$$V^i(k, x) = \max_{u_k^i}\left\{\left[P_k^i u_k^i - \frac{c_k^i}{x_k}(u_k^i)^2\right]\delta_1^k \right.$$

$$\left. + \sum_{y=1}^{4}\gamma_k^y V^i[k+1, x + a - bx - u_k^i - u_k^{j**} + \vartheta_k^y x_k]\right\} \tag{29}$$

for $i \in \{1, 2\}$, $i \neq j$, *and* $k \in \{1, 2, \ldots, T\}$.

Solving Corollary 3.1 yields the following.

Proposition 3.1. *The value function indicating the expected game equilibrium payoff of player* $\boldsymbol{i}$ *is*

$$V^i(k, x) = \left(A_k^i x + C_k^i\right)\delta_1^k, \quad \text{for } i \in \{1, 2\} \quad \text{and } k \in \{1, 2, \ldots, T\}, \tag{30}$$

where

$$\boldsymbol{A^i_{T+1} = q^i_{T+1}, C^i_{T+1} = Q^i_{T+1}};$$

$$A^i_k = \frac{1}{4c^i_k}\left(P^i_k + \sum_{y=1}^{4}\gamma^y_k A^i_{k+1}\delta^{k+1}_k\right)\left(P^i_k - \sum_{y=1}^{4}\gamma^y_k A^i_{k+1}\delta^{k+1}_k\right)$$
$$+ A^i_{k+1}\left[\sum_{h=1}^{4}\vartheta^h_k - b - \sum_{j=1}^{2}\frac{x}{2c^j_k}\left(P^j_k - \sum_{y=1}^{4}\gamma^y_k A^j_{k+1}\delta^{k+1}_k\right)\right]\delta^{k+1}_k,$$
$$C^i_k = \left(A^i_{k+1}a + C^i_{k+1}\right)\delta^{k+1}_k.$$

The game equilibrium strategies can be obtained as

$$u^{i**}_k = \frac{x}{2c^i_k}\left(P^i_k - \sum_{y=1}^{4}\gamma^y_k A^i_{k+1}\delta^{k+1}_k\right) \tag{31}$$

for $i \in \{1, 2\}$ *and* $k \in \{1, 2, \ldots, T\}$.

Proof. See Appendix A. □

If the extractors decide to cooperate, they first have to solve the following stochastic optimization to maximize their joint profit:

$$\boldsymbol{\max_{\substack{u^1_k, u^2_k\\ k\in\{1,2,\ldots,T\}}} E_{\vartheta_1,\vartheta_2,\ldots,\vartheta_T}\left\{\sum\nolimits_{j=1}^{2}\left[\sum\nolimits_{k=1}^{T}\left(P^j_k u^j_k - \frac{c^j_k}{x_k}(u^j_k)^2\right)\delta^k_1\right.\right.}$$
$$\left.\left.+\sum\nolimits_{\mathbf{j}=1}^{2}\left(\boldsymbol{q^j_{T+1}x_{T+1} + Q^j_{T+1}}\right)\boldsymbol{\delta^{T+1}_1}\right]\right\} \tag{32}$$

subject to dynamics (26).

Invoking Theorem 1.2, one can characterize the optimal controls in the stochastic dynamic programming problem (26) and (32) in the following corollary.

Corollary 3.2. *A set of strategies* $\{u^{i*}_k = \psi^i_k(x)$, *for* $k \in \{1, 2, \ldots, T\}$ *and* $i \in \{1, 2\}\}$ *provides an optimal solution to the stochastic dynamic programming problem (26) and (32) if there exist* *functions $W^N(k,x)$, for $k \in \{1,2,\ldots,T\}$, such that the following recursive relations are satisfied:*

$$W^N(T+1,x) = \sum_{i=1}^{2} (q^i_{T+1}x_{T+1} + Q^i_{T+1})\delta_1^{T+1}, \tag{33}$$

$$W^N(k,x) = \max_{u_k^1,u_k^2} \left\{ \sum_{j=1}^{2} \left(P_k^j u_k^j - \frac{c_k^j}{x_k}(u_k^j)^2 \right) \delta_1^k \right.$$

$$\left. + \sum_{y=1}^{4} \gamma_k^y W^N[k+1, x+a-bx-u_k^1-u_k^2+\vartheta_k^y x_k] \right\} \tag{34}$$

for $i \in \{1,2\}$, $i \neq j$, and $k \in \{1,2,\ldots,T\}$.

Solving Corollary 3.2 yields the following.

Proposition 3.2. *The value function indicating the expected joint profit of the extractors under cooperation is*

$$\boldsymbol{W^N(k,x) = (A_k x + C_k)\delta_1^k}, \quad \text{for } \boldsymbol{k \in \{1,2,\ldots,T\}}, \tag{35}$$

where

$$\boldsymbol{A_{T+1} = \sum_{j=1}^{2} q^j_{T+1}}, \quad \boldsymbol{C_{T+1} = \sum_{j=1}^{2} \quad Q^j_{T+1}};$$

$$A_k = \sum_{j=1}^{2} \frac{1}{4c_k^j} \left(P_k^j + \sum_{y=1}^{4} \gamma_k^y A_{k+1}\delta_k^{k+1} \right) \left(P_k^j - \sum_{y=1}^{4} \gamma_k^y A_{k+1}\delta_k^{k+1} \right)$$

$$+ A_{k+1} \left[\sum_{h=1}^{4} \vartheta_k^h - b - \sum_{j=1}^{2} \frac{x}{2c_k^j} \left(P_k^j - \sum_{y=1}^{4} \gamma_k^y A_{k+1}\delta_k^{k+1} \right) \right] \delta_k^{k+1},$$

$$C_k = (A_{k+1}a + C_{k+1})\,\delta_k^{k+1}.$$

The optimal cooperative strategies can be obtained as

$$u_k^{i*} = \frac{x}{2c_k^i} \left(P_k^i - \sum_{y=1}^{4} \gamma_k^y A_{k+1}\delta_k^{k+1} \right) \tag{36}$$

for $i \in \{1,2\}$ and $k \in \{1,2,\ldots,T\}$.

Proof. See Appendix B. □

Substituting the optimal strategies in (36) into (26) yields the dynamics of the optimal state:

$$x_{k+1} = x_k + a - bx_k - \sum_{i=1}^{2} \frac{x_k}{2c_k^i} \left(P_k^i - \sum_{y=1}^{4} \gamma_k^y A_{k+1} \delta_k^{k+1} \right) + \vartheta_k x_k. \tag{37}$$

We use X_k^* to denote the set of realizable values of x_k^* at stage k generated by (38). The term $x_k^* \in X_k^*$ is used to denote an element in X_k^*.

To obtain the marginal contributions of the player, we have to solve the game:

$$\max_{\substack{u_k^i \\ k \in \{1,2,\ldots,T\}}} \quad \boldsymbol{E}_{\vartheta_1,\vartheta_2,\ldots,\vartheta_T} \left\{ \sum_{k=1}^{T} \left(P_k^i u_k^i - \frac{c_k^i}{x_k} (u_k^i)^2 \right) \delta_1^k \right.$$
$$\left. + (q_{T+1}^i x_{T+1} + Q_{T+1}^i) \delta_1^{T+1} \Big] \right\}, \quad \text{for } i \in \{1, 2\} \tag{38}$$

subject to dynamics (26).

Since there are only two extractors, $W^{N \backslash 2}(k, x) = V^1(k, x)$ and $W^{N \backslash 1}(k, x) = V^2(k, x)$.

A dynamically stable fair gain-sharing solution can be obtained as

$$\xi^i\left(k, x_k^*\right) = \frac{W^N\left(k, x_k^*\right) - V^j\left(k, x_k^*\right)}{\sum_{\zeta=1}^2 \left[W^N\left(k, x_k^*\right) - V^\zeta\left(k, x_k^*\right)\right]}$$
$$\times \left[W^N\left(k, x_k^*\right) - \sum_{j=1}^{2} V^j\left(k, x_k^*\right)\right] + V^i\left(k, x_k^*\right),$$
$$= \frac{A_k x + C_k - \left(A_k^j x + C_k^j\right)}{\sum_{\zeta=1}^2 \left[A_k x + C_k - \left(A_k^\zeta x + C_k^\zeta\right)\right]}$$

$$\times \left[A_k x + C_k - \sum_{\zeta=1}^{2} \left(A_k^{\zeta} x + C_k^{\zeta} \right) \right] \delta_1^k$$
$$+ (A_k^i x + C_k^i)\delta_1^k, \text{ for } k \in \{1, 2, \ldots, T\} \text{ and } \{i \in 1, 2\}. \quad (39)$$

Following the analysis in Section 2, we let $B_k^i(x_k^*)$ denote the payment that player i will receive at stage k under the cooperative agreement in (39). The payment scheme $B_k^i(x_k^*)$ constitutes a PDP in the sense that the imputation to player i over the stages k to $T+1$ can be expressed as

$$\xi^i (k, x_k^*) = B_k^i (x_k^*) \delta_1^k + E_{\vartheta_{k+1}, \vartheta_{k+2}, \ldots\cdots, \vartheta_T} \left(\sum_{\zeta=k+1}^{T} B_\zeta^i (x_\zeta^*) \delta_1^\xi + q^i (x_{T+1}^*) \delta_1^{T+1} \right) \quad (40)$$

for $i \in \{1, 2\}$.

Invoking Theorem 2.2, a payment equaling

$$B_k^i(x_k^*) = \frac{1}{\delta_1^k} \left[\xi^i (k, x_k^*) - \sum_{y=1}^{4} \gamma_k^y \xi^i [k+1, x + a - bx - u_k^{1*} - u_k^{2*} + \vartheta_k^y x_k^*] \right] \quad (41)$$

given to player $i \in N$ at stage $k \in \{1, 2, \ldots, T\}$ would lead to the realization of the imputation in (39).

4. A Note on Uncertain Future Payoff Structures

In strategic decision-making, whether in economics, politics, social, military, or scientific games, players face payoff structures that are rarely fixed or predictable. The sources of uncertainty surrounding future payoffs include (i) incomplete information, (ii) dynamic environments, (iii) the interdependence of choices, and (vi) external random shocks. Identifying, understanding, and managing these uncertainties on the payoff structures is crucial for anticipating the players' strategies.

(i) **Incomplete Information:** Players often lack full knowledge of opponents' strategies, preferences, or resources. This insufficiency creates uncertainty in evaluating future payoff structures and predicting the game outcomes.
(ii) **Dynamic Environments:** Payoff structures evolve over time due to technological innovation, regulatory changes, or shifting social norms. What appears optimal today may become suboptimal tomorrow.
(iii) **Interdependence of Choices:** In multi-player settings, payoffs depend not only on one's own actions but also on the unpredictable responses of others. Players may form coalitions, joint ventures, and collaborations that affect the payoff structures in the future.
(iv) **External Shocks:** Events such as economic crises, geopolitical conflicts, unexpected scientific advancements, or natural disasters can radically alter payoff structures, rendering prior strategies ineffective.

The uncertain future payoff structures of players highlight the complexity of strategic interaction in real-world contexts. Rather than seeking perfect foresight, players must embrace uncertainty as an inherent feature of decision-making. Adaptability and the capacity to anticipate multiple possible payoff structures are necessary for practical operation.

To highlight the random phenomenon in the payoff structures explicitly, we consider the evolution dynamics:

$$x_{k+1} = f_k(x_k, u_k^1, u_k^2, \ldots, u_k^n) + \vartheta_k G(x_k), \quad x_1 = x_1^0, \tag{42}$$

where $x_k = (x_k^1, x_k^2)$.

In particular, x_k^1 is a vector of random variables related to the structure of the payoff functions. The expected payoff of player i

from stage 1 to stage T can be expressed as

$$E_{\vartheta_1,\vartheta_2,\ldots,\vartheta_T}\left\{\sum_{k=1}^{T} g_k^i(x_k^1, x_k^2, u_k^1, u_k^2, \ldots, u_k^n)\delta_1^k + q^i(x_{T+1})\delta_1^{T+1}\right\}, \tag{43}$$

for $i \in N$.

5. Concluding Remarks

This chapter explores cooperative solutions that guarantee fair gain-sharing and maintain time consistency in the face of uncertainty. It discusses both stochastic dynamic games, which account for uncertainty in the progression of state dynamics, and randomly furcating stochastic dynamic games, which introduce uncertainties in the players' payoff structures as well as in the progression of state dynamics. The notion of randomly furcating stochastic dynamic games was introduced by Yeung and Petrosyan (2013). The shift from continuous-time randomly furcating stochastic differential games to a discrete-time framework is not only of theoretical importance but also carries practical implications for applications in operations research. In the process of deriving the main results for subgame-consistent solutions, Nash equilibria for stochastic dynamic games and optimal control strategies for randomly furcating stochastic control problems are also established. Yeung and Petrosyan (2014a) investigated the subgame-consistent cooperative provision of public goods in the context of accumulation and payoff uncertainties. Additionally, Yeung and Petrosyan (2014b) examined a subgame-consistent solution for a dynamic game centered on pollution management, where future environmental costs remain uncertain. The initial examination of subgame-consistent solutions for cooperative stochastic dynamic games was conducted by Yeung and Petrosyan (2010). Subgame-consistent cooperative solutions for randomly furcating stochastic dynamic games can be found in the works of Petrosyan and Yeung (2007), and Yeung and Petrosyan (2013, 2014c). The application of stochastic differential games in transboundary industrial pollution was presented by Yeung and Petrosyan (2008). Yeung and Petrosyan (2026) analyzed cooperative dynamic games and artificial intelligence. Yeung and Petrosyan (2006) co-authored the first book on cooperative stochastic differential games. Furthermore, Yeung and Petrosyan (2016) co-authored a detailed treatise on subgame-consistent cooperation, which encompassed cooperative stochastic dynamic games and randomly furcating stochastic dynamic games in both continuous time and discrete time.

Appendices

Appendix A. Proof of Proposition 3.1

Using the first equation in Corollary 3.1, we obtain

$$\begin{aligned} V^i(T+1,x) &= \left(q^i_{T+1}x_{T+1} + Q^i_{T+1}\right)\delta_1^{T+1} \\ &\equiv \left(A^i_{T+1}x_{T+1} + C^i_{T+1}\right)\delta_1^{T+1}. \end{aligned} \tag{A.1}$$

Invoking the technique of backward induction, we consider first the last stage of operation, i.e., stage T. Performing the indicated maximization operator in Corollary 3.1, we can obtain the game equilibrium strategies in stage T as

$$\begin{aligned} \boldsymbol{u}_{\boldsymbol{T}}^{\boldsymbol{i}**} &= \left(\frac{1}{\delta_1^{\mathrm{T}}}\right)\frac{\mathrm{x}}{\mathbf{2}\boldsymbol{c}_{\boldsymbol{T}}^{\mathrm{i}}}\left(\boldsymbol{P}_{\boldsymbol{T}}^{\boldsymbol{i}}\right. \\ &\quad \left. -\sum_{\boldsymbol{y}=\mathbf{1}}^{\mathbf{4}}\gamma_{\boldsymbol{k}}^{\boldsymbol{y}}\boldsymbol{V}^{\boldsymbol{i}}_{\boldsymbol{x}_{\boldsymbol{T}+\mathbf{1}}}\left[\boldsymbol{T}+\mathbf{1}\boldsymbol{x}+\boldsymbol{a}-\boldsymbol{b}\boldsymbol{x}-\boldsymbol{u}_{\boldsymbol{T}}^{\boldsymbol{i}**}-\boldsymbol{u}_{\boldsymbol{T}}^{\mathbf{2}**}+\vartheta_{\boldsymbol{T}}^{\boldsymbol{y}}\boldsymbol{x}\right]\right) \end{aligned} \tag{A.2}$$

for $i \in \{1,2\}$ and $i \neq j$.

Invoking (A.1), we obtain

$$\boldsymbol{u}_{\boldsymbol{T}}^{\boldsymbol{i}**} = \frac{\boldsymbol{x}}{\mathbf{2}\boldsymbol{c}_{\boldsymbol{T}}^{\boldsymbol{i}}}\left(\boldsymbol{P}_{\boldsymbol{T}}^{\boldsymbol{i}} - \sum_{\boldsymbol{y}=\mathbf{1}}^{\mathbf{4}}\gamma_{\boldsymbol{T}}^{\boldsymbol{y}}\boldsymbol{A}_{\boldsymbol{T}+\mathbf{1}}^{\boldsymbol{i}}\boldsymbol{\delta}_{\boldsymbol{T}}^{\boldsymbol{T}+\mathbf{1}}\right) \tag{A.3}$$

Substituting (A.3) into the equation for stage T in Corollary 3.1 yields

$$\begin{aligned}V^i(T,x) &= \frac{x}{4c_T^i}\left(P_T^i + \sum_{y=1}^{4}\gamma_T^y A_{T+1}^i \delta_T^{T+1}\right) \\ &\quad \times \left(P_T^i - \sum_{y=1}^{4}\gamma_k^y A_{T+1}^i \delta_T^{T+1}\right)\delta_1^T \\ &\quad + \left\{A_{T+1}^i \left[a - bx - \sum_{j=1}^{2}\frac{x}{2c_T^j}\left(P_T^j - \sum_{y=1}^{4}\gamma_k^y A_{T+1}^j \delta_T^{T+1}\right)\right.\right. \\ &\quad \left.\left. + \sum_{h=1}^{4}\vartheta_T^h x\right] + C_{T+1}^i\right\}\delta_1^{T+1} \quad \text{for } i \in \{1,2\}. \qquad \text{(A.4)}\end{aligned}$$

Since the right-hand side of the equation system (A.4) has functions linear in x, $V^i(Tx)$ can be expressed as $(A_T^i x + C_T^i)\delta_1^T$, where

$$\begin{aligned}A_T^i &= \frac{1}{4c_T^i}\left(P_T^i + \sum_{y=1}^{4}\gamma_T^y A_{T+1}^i \delta_T^{T+1}\right)\left(P_T^i - \sum_{y=1}^{4}\gamma_T^y A_{T+1}^i \delta_T^{T+1}\right) \\ &\quad + A_{T+1}^i\left[\sum_{h=1}^{4}\vartheta_T^h - b - \sum_{j=1}^{2}\frac{1}{2c_T^j}\left(P_T^j - \sum_{y=1}^{4}\gamma_T^y A_{T+1}^j \delta_T^{T+1}\right)\right]\delta_T^{T+1}, \\ C_T^i &= \left(A_{T+1}^i a + C_{T+1}^i\right)\delta_T^{T+1}. \qquad \text{(A.5)}\end{aligned}$$

Now, we consider the cases for stage $k \in \{T-1, T-2, \ldots, 1\}$. Repeating the analysis in (A.1)–(A.5) backward sequentially from stage $k = T-1$ to stage $k = 1$, we obtain

$$\boldsymbol{u_k^{i**} = \frac{x}{2c_k^i}\left(P_k^i - \sum_{y=1}^{4}\gamma_k^y A_{k+1}^i \delta_k^{k+1}\right)} \quad \text{for } \boldsymbol{i \in \{1,2\}}. \qquad \text{(A.6)}$$

Substituting (A.6) into the equation for stage k in Corollary 3.1 yields

$$\begin{aligned} V^i(k,x) = {} & \frac{x}{4c_k^i}\left(P_k^i + \sum_{y=1}^{4}\gamma_k^y A_{k+1}^i \delta_k^{k+1}\right) \\ & \times \left(P_k^i - \sum_{y=1}^{4}\gamma_k^y A_{k+1}^i \delta_k^{k+1}\right)\delta_1^k \\ & + \left\{A_{k+1}^i\left[a - bx - \sum_{j=1}^{2}\frac{x}{2c_k^j}\left(P_k^j - \sum_{y=1}^{4}\gamma_k^y A_{k+1}^j \delta_k^{T+1}\right)\right.\right. \\ & \left.\left. + \sum_{h=1}^{4}\vartheta_k^h x\right] + C_{k+1}^i\right\}\delta_1^{k+1} \end{aligned} \tag{A.7}$$

for $i \in \{1,2\}$.

Since the right-hand side of the equation system (A.7) has functions linear in x, $V^i(k,x)$ can be expressed as $(A_k^i x + C_k^i)\delta_1^k$, where

$$\begin{aligned} A_k^i = {} & \frac{1}{4c_k^i}\left(P_k^i + \sum_{y=1}^{4}\gamma_k^y A_{k+1}^i \delta_k^{k+1}\right)\left(P_k^i - \sum_{y=1}^{4}\gamma_k^y A_{k+1}^i \delta_k^{k+1}\right) \\ & + A_{k+1}^i\left[\sum_{h=1}^{4}\vartheta_k^h - b - \sum_{j=1}^{2}\frac{1}{2c_k^j}\left(P_k^j - \sum_{y=1}^{4}\gamma_k^y A_{k+1}^j \delta_k^{k+1}\right)\right]\delta_k^{k+1}, \\ C_k^i = {} & \left(A_{k+1}^i a + C_{k+1}^i\right)\delta_k^{k+1}. \end{aligned} \tag{A.8}$$

Hence, Proposition 3.1 follows.

Appendix B. Proof of Proposition 3.2

Using the first equation in Corollary 3.2, we obtain

$$\begin{aligned} W^N(T+1,x) &= \sum_{i=1}^{2}(q_{T+1}^i x_{T+1} + Q_{T+1}^i)\delta_1^{T+1} \\ &\equiv (A_{T+1}x_{T+1} + C_{T+1})\delta_1^{T+1}. \end{aligned} \tag{B.1}$$

Invoking the technique of backward induction, we consider first the last stage of operation, i.e., stage T. Performing the indicated maximization operator in Corollary 3.2, we can express the optimal cooperative strategies in stage T as

$$\boldsymbol{u_T^{i*} = \left(\frac{1}{\delta_1^T}\right)\frac{x}{2c_T^i}\left(P_T^i - \sum_{y=1}^{4}\gamma_T^y W_{x_{T+1}}^N[T+1, x+a-bx - u_T^{i*} - u_T^{2*} + \vartheta_T^y x]\right)} \tag{B.2}$$

for $i \in \{1,2\}$ and $i \neq j$.

Invoking (B.1), we obtain

$$\boldsymbol{u_T^{i**} = \frac{x}{2c_T^i}\left(P_T^i - \sum_{y=1}^{4}\gamma_T^y A_{T+1}\delta_T^{T+1}\right)}. \tag{B.3}$$

Substituting (B.3) into the equation for stage T in Corollary 3.2 yields

$$\begin{aligned} W^N(T,x) = {} & \sum_{j=1}^{2}\frac{x}{4c_T^j}\left(P_T^i + \sum_{y=1}^{4}\gamma_T^y A_{T+1}\delta_T^{T+1}\right) \\ & \times \left(P_T^j - \sum_{y=1}^{4}\gamma_T^y A_{T+1}\delta_T^{T+1}\right)\delta_1^T \\ & + \left\{A_{T+1}\left[a - bx - \sum_{j=1}^{2}\frac{x}{2c_T^j}\left(P_T^j - \sum_{y=1}^{4}\gamma_T^y A_{T+1}^j\delta_T^{T+1}\right)\right.\right. \\ & \left.\left. + \sum_{h=1}^{4}\vartheta_T^h x\right] + C_{T+1}\right\}\delta_1^{T+1}. \end{aligned} \tag{B.4}$$

Since the right-hand side of equation (B.4) is a function linear in x, $W^N(T,x)$ can be expressed as $(A_T x + C_T)\delta_1^T$, where

$$A_T = \sum_{j=1}^{2}\frac{1}{4c_T^j}\left(P_T^i + \sum_{y=1}^{4}\gamma_T^y A_{T+1}\delta_T^{T+1}\right) \times \left(P_T^j - \sum_{y=1}^{4}\gamma_T^y A_{T+1}\delta_T^{T+1}\right) + A_{T+1}\left[\sum_{h=1}^{4}\vartheta_T^h - b - \sum_{j=1}^{2}\frac{1}{2c_T^j}\right.$$

$$\times \left(P_T^j - \sum_{y=1}^{4} \gamma_T^y A_{T+1} \delta_T^{T+1} \right) \Bigg] \delta_T^{T+1},$$

$$C_T = (A_{T+1} a + C_{T+1})\, \delta_T^{T+1}. \tag{B.5}$$

Now, we consider the cases for stage $k \in \{T-1,\ T-2, \ldots, 1\}$. Repeating the analysis in (A.1)–(A.5) backward sequentially from stage $k = T-1$ to stage $k = 1$, we obtain

$$\boldsymbol{u_k^{i*} = \frac{x}{2c_k^i} \left(P_k^i - \sum_{y=1}^{4} \gamma_k^y A_{k+1} \delta_k^{k+1} \right) \quad \text{for } i \in \{1, 2\}}. \tag{B.6}$$

Substituting (B.6) into the equation for stage k in Corollary 3.2 yields

$$\begin{aligned} W^N(k,x) &= \sum_{j=1}^{2} \frac{x}{4c_k^j} \left(P_k^i + \sum_{y=1}^{4} \gamma_k^y A_{k+1} \delta_k^{k+1} \right) \\ &\quad \times \left(P_k^j - \sum_{y=1}^{4} \gamma_k^y A_{k+1} \delta_k^{k+1} \right) \delta_1^{k+1} \\ &\quad + \left\{ A_{k+1} \left[a - bx - \sum_{j=1}^{2} \frac{x}{2c_k^j} \left(P_k^j - \sum_{y=1}^{4} \gamma_k^y A_{k+1} \delta_k^{T+1} \right) \right. \right. \\ &\quad \left. \left. + \sum_{h=1}^{4} \vartheta_k^h x \right] + C_{k+1} \right\} \delta_1^{k+1}. \end{aligned} \tag{B.7}$$

Since the right-hand side of equation (B.7) has a function linear in x, $W^N(k,x)$ can be expressed as $(A_k x + C_k)\delta_1^k$, where

$$\begin{aligned} A_k &= \sum_{j=1}^{2} \frac{1}{4c_k^j} \left(P_k^J + \sum_{y=1}^{4} \gamma_k^y A_{k+1} \delta_k^{k+1} \right) \left(P_k^j - \sum_{y=1}^{4} \gamma_k^y A_{k+1}^j \delta_k^{k+1} \right) \\ &\quad + A_{k+1} \left[\sum_{h=1}^{4} \vartheta_k^h - b - \sum_{j=1}^{2} \frac{1}{2c_k^j} \left(P_k^j - \sum_{y=1}^{4} \gamma_k^y A_{k+1} \delta_k^{k+1} \right) \right] \delta_k^{k+1}, \end{aligned}$$

$$C_k = (A_{k+1} a + C_{k+1})\, \delta_k^{k+1}. \tag{B.8}$$

Hence, Proposition 3.2 follows.

References

Petrosyan, L. A., & Yeung, D. W. K. (2007). Subgame-consistent cooperative solutions in randomly furcating stochastic differential games. *Mathematical and computer modelling International Journal of Mathematical and Computer Modelling (Special Issue on Lyapunov's Methods in Stability and Control)*, 45(11-12), 1294–1307.

Yeung, D. W. K., & Petrosyan, L. A. (2006). *Cooperative Stochastic Differential Games*. New York, NY: Springer.

Yeung, D. W. K., & Petrosyan, L. A. (2008). A cooperative stochastic differential game of transboundary industrial pollution. *Automatica*, 44(6), 1532–1544.

Yeung, D. W. K., & Petrosyan, L. A. (2010). Subgame consistent solutions for cooperative stochastic dynamic games. *Journal of Optimization Theory and Applications*, 145(1), 579–596.

Yeung, D. W. K., & Petrosyan, L. A. (2011). Subgame consistent cooperative solution of dynamic games with random horizon. *Journal of Optimization Theory and Applications*, 150(1), 78–97.

Yeung, D. W. K., & Petrosyan, L. A. (2012). *Subgame Consistent Economic Optimization: An Advanced Cooperative Dynamic Game Analysis*. Boston: Birkhäuser. ISBN 978-0-8176-8261-3, p. 395.

Yeung, D. W. K., & Petrosyan, L. A. (2013). Subgame-consistent cooperative solutions in randomly furcating stochastic dynamic games. *Mathematical and Computer Modelling*, 57(3-4), 976–991.

Yeung, D. W. K., & Petrosyan, L. A. (2014a). Subgame consistent cooperative provision of public goods under accumulation and payoff uncertainties. In J. Haunschmied, V. M. Veliov, & S. Wrzaczek (eds.), *Dynamic Games in Economics*, pp. 289–315. Berlin Heidelberg: Springer-Verlag.

Yeung, D. W. K., & Petrosyan, L. A. (2014b). Subgame consistent solution for a dynamic game of pollution management. In L. A. Petrosyan, J. V. Romanovsky, & D. W. K. Yeung (eds.): *Advances in Economics and Optimization: Collected Scientific Studies Dedicated to the Memory of L.V. Kantorovich*, New Nova Science Publishers, 2014b, pp. 95–112.

Yeung, D. W. K., & Petrosyan, L. A. (2014c). Subgame consistent cooperative solutions for randomly furcating stochastic dynamic games with uncertain horizon. *International Game Theory Review*, 16(02), 1440012.

Yeung, D. W. K., & Petrosyan, L. A. (2016). *Subgame Consistent Cooperation. A Comprehensive Treatise.* Singapore: Springer, p. 520, ISBN 987-981-10-1544-1.

Yeung, D. W. K., & Petrosyan, L. A. (2026). Cooperative Dynamic Games and Artificial Intelligence: A Theoretical Perspective, forthcoming in M. Rassias (ed.) *Game Theory and Artificial Intelligence*, Switzerland: Springer, 2026.

Index

A

advertising, 176
air pollution, 177
asymmetry, 5, 9, 12–14, 17, 49, 64, 76, 105, 148, 180

B

Bellman equation, 123–124, 139
blocs, 58, 162

C

clean technology, 12, 49, 60, 69–70
climate change, 51, 53–55, 88
coalition, 3, 11, 35, 57, 75, 182
commitment, 20, 57, 67
Conference of the Parties (COP), 48, 51–52, 54–59, 61, 67–68
cooperative dynamic game, 5, 17, 34, 139
cooperative games, 9, 139
cooperative solution, 3, 5–7, 18, 20, 28, 48, 64–65, 68–69, 80, 89, 93, 95, 155, 175–176, 180, 192

D

differential games, 139
diffusion, 2
dynamic cooperation, 9, 88
dynamic game, 6, 8, 14–15, 27, 70, 90, 102, 117, 119, 148, 177, 180, 192
dynamic network game, 77, 149
dynamic optimization, 68, 91, 150
dynamic programming, 6, 15, 90–91, 95, 139, 151, 153, 179
dynamically consistent solutions, 139

E

ecosystem, 61–62, 68
efficient collaboration, 7, 147
emissions, 52–54, 56, 68
energy, 56–57, 60, 62, 68
environmental degradation, 59
environmental management, 69, 175
equal sharing of cooperative gains, 10

F

failure on climate finance, 59
feedback Nash equilibrium, 15, 27, 63, 77, 90, 123, 152, 178, 186

G

global climate agreement, 54
global environmental management, 4–5, 47
global environmental pact, 13
globalization, 1–3, 5–6, 47, 85–86
green technology, 61–62, 69

green technology development, 70
group optimal outcome, 16, 61, 71
group optimality, 4–5, 7, 22, 25, 34, 77, 88, 96, 105, 123–124, 154, 162, 179
Guterres, António, 3

H

Hamilton–Jacobi–Bellman (HJB) equations, 90, 94

I

imputation, 21, 129–131, 139, 183–184, 190
incentives, 60, 129
individual rationality, 1, 4–5, 7, 10, 23, 25, 47, 97, 147, 154, 162
innovation, 60–61, 68, 105–106, 176
interconnected electricity network, 148
interconnectivity of electricity generation, 13
international cooperation, 1

J

joint venture, 12, 85–86, 89, 92, 101, 106

L

life span, 6

M

marginal contributions, 7, 18, 65, 86, 93, 95, 104, 147, 175, 180, 183, 189

N

Nash equilibrium, 19, 63, 77, 90, 95, 123–124, 128, 133, 153, 178, 182, 186
networks, 4, 7, 14, 86, 162

O

objectives of the cooperation scheme, 62
optimal cooperative trajectory, 129
optimality principle, 7, 14, 17, 25, 124, 147, 149

P

Paris agreement, 56, 58
payment mechanism, 129
payoff distribution procedure (PDP), 23, 95, 97, 130–131, 147, 153, 155, 161, 183–184, 190
policy instruments, 59
pollution accumulation, 69

R

random exit, 7, 117–118, 132, 137
random variable, 119, 139, 177–178, 185
randomly furcating stochastic dynamic games, 192–193
remedial solution, 5, 47
resolution plan, 67
resource extraction, 132, 185

S

Shapley value, 11, 35, 139
spillover, 87, 100, 148
state dynamics, 17, 64, 72, 119, 192
stochastic differential games, 192
stochastic dynamic game, 7, 175–176, 181, 186
subgame consistency, 129, 131
subgame-consistent, 23, 118, 129, 139, 180, 183–184, 192

T

target controls, 37
terminal payoff, 15, 70, 89
termination, 6
time consistency, 4, 10, 20–22, 25, 34, 47, 77, 88, 96, 147–148, 154, 162, 192

U

uncertainties, 6, 86, 117, 192

www.ingramcontent.com/pod-product-compliance
Lightning Source LLC
LaVergne TN
LVHW051051210926
848486LV00002B/73